AF893868

Nataraj Books
7073 Brookfield Plaza
Springfield, VA 22150
Phone: (703) 455-4996

ENERGY OPTIONS FOR 21ST CENTURY

Energy Options for 21st Century

R.N. Subudhi

ASHISH PUBLISHING HOUSE
8/81, Punjabi Bagh, New Delhi-110 026

Published by:
S.B. Nangia
for **Ashish Publishing House**
8/81, Punjabi Bagh,
New Delhi-110 026
Tel.: 500581, 5410924

Showroom : 5, Ansari Road, Delhi 110 002
Tel : 3274050

ISBN 81-7024-538-9

1993

© R.N. Subudhi

This edition is copyright under the Berne Convention. All rights are reserved. Apart from any fair dealing for the purpose of private study, research, criticism or review as permitted under the Copyright Act, 1956, no part of this publication may be reproduced, stored in a retrieval system, or transmitted, in any form or by any means, whatsoever, without the prior permission of the copyright owner. Enquiries should be addressed to M/s Ashish Publishing House, 8/81, Punjabi Bagh, New Delhi-110 026.

Laser Typeset by:
Westvill Enterprises
Block B-5, Bungalow No. 47
Paschim Vihar,
New Delhi-110 063
Tel.: 558-7469

Printed at :
EFFICIENT OFFSET PRINTERS
215, Shahzada Bagh Indl. Complex,
Phase-II, Phone : 533736, 533762
Delhi - 110035

Preface

When we talk of the energy-future for a country like India, with a fast exploding population problem, all the forecasting techniques appear to bite dust. In terms of population, a 'Second India' is in the offing within a few years. By the turn of the country, it is estimated, over 530 million tonnes of Coal and 92 million tonnes of oil would be required per annum for the nation. It is also estimated (by the Planning Commission's working group) that, by the year 2004-05, over 500 billion units (Kwh) of electrical energy would be required, with a total installed capacity of at least 1,50,000 MW. (Presently we produce about 150 million units of Electricity with a generation capacity of 43000 MW only).

Problem and Mission

It is not the "exhorbitantly a very high cost" of generation (Eighth Plan estimate puts it at Rs.107,000 crore for the additional generation of 31,000 MW), that worries the planners now; our environment, and of course the human life, is at stake because of the dangerous side effects of our energy-system and hazardous wastes of Power-generating units. Be it the Earth Summit or a meet on Global Warming; the principal concern now is one. For the progress of the human race (which in the process depletes natural resources), we can afford to think of a "SUSTAINABLE' development, that must take care of the needs of subsequent generations and that must take care of our environment, our own existence too.

We of course need badly more power; but to think of safer means now and to think of optimum use of the existing resource. Our present attempt, in this line, is to explore and suggest some better possibilities for the future. Starting from energy planning to energy use, through generation and distribution, for every stage, equal attention is called for. Problem is both complex and diverse. It requires a series of works.

Electricity Sector

Development process of a country is directly dependent on the continuous support of the energy sector of that country. The fact that, energy is never available in abundance, leaves before us the only option of managing properly the limited available energy for is optimum use. According to the latest Planning Commission report, only SEBs account for 96% of total electricity sold in India. Keeping in view the vast population size and the increasing energy utilisation need, the management and operational problem of the massive network of SEBs has now become more tough these days. Though SEB's. as per their statute, are supposed to make a profit of atleast 3 per cent of the total investment, all most every SEB is making loss continuously.

Despite the promulgation of special *Indian Electricity Act (1910)* and recent national concern expressed through the report of Committee on Power (*1980*), the malady still pervades. The average *transmission and distribution (T&D)* losses for India is rated over 21%, with a maximum individual properation of loss being as high as 49.5% for J&K. Where as for the developed countries, this rate is less than 10%. In absolute terms, UPSEB tops the list with a total T&D loss of 5.96 billion kwh. A large amount of energy is thus lost at a time when the very SEBs borrow energy from other agencies like central electricity undertakings and state corporations. This external purchase is almost 32% of the total energy availability. There is not only the need, also there are scopes too to bring down the existing T&D losses through improved technical and administrative measures. Contracting and sub-contracting some job, as is the practice with some SEBs, is not fitting solution. Theft of energy, the measure cause of the loss, is now being treated as a cognizible offence. However, the cause of the higher amount of loss is, since the past, due to the continuation of inefficient T&D system/equipments.

According to the report, at present the cost of estblishment and administartion is 16.8 paise per kwh and cost of operation and maintenance is 5.2 paise per kwh sold. The average rate for sale of energy is 75.1 paise per kwh sold. But because of Govt. policy, the rate is very low for agriculture and domestic sectors. The rate of revenue from agricultural sector is as low as 15.3 paise per kwh sold. Exhibit-1 shows the quantum of loss on these sectors.

Exhibit - 1 (Estimated loss of all SEBs)

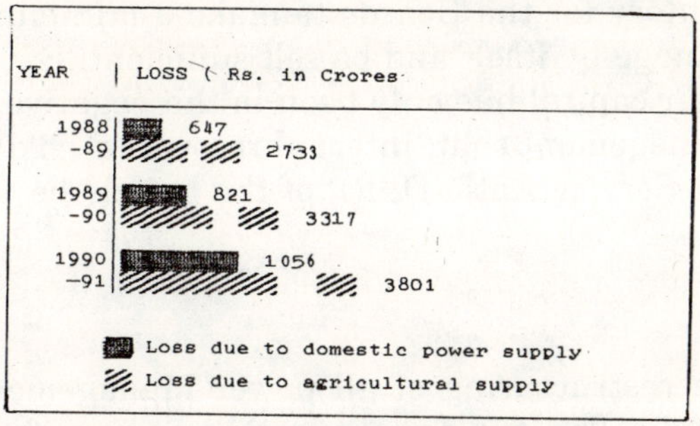

Boards realise their revenue mainly from the energy sold. Ofcourse SEBs get Govt. subsidies for the loss due to such populist policies. But realisaion part of the arrears, is altogether a different story. What to talk of the normal time-lag and the running arrears, beaclog arrears have been found to be accumulating hopelessly. Total arrear has increased from Rs.1345 crores at end of 1984-85 to Rs.2964 crores at the end of 1988-89. This amount is equivalent to 3 months total revenue ! Such long non-realisation badly strains the "the cash liquidity of Boards impairing in the process, the ability of the boards meeting their commitments to their suppliers of fuel, imported power etc."

The rate of growth in manpower is meagre in comparision to the increasing workload. While energy distribution and resulting management-load has increased in a geometric Progression, employment strength has swelled only in an arithmetic fashion. Again, when we compare with the amount of energy sold and the propertion of customers, the effective strength has come down significantly. This evident is from Exhibit -II.

Exhibit - II (Man power position of SEBs)

Year	*Total Staff*	*No. of Employees per*	
	('000 No.Rounded)	*MKWH Sold*	*1000 Consumers*
1985-86	842	7.3	20.1
1986-87	852	6.7	18.9
1987-88	872	6.4	17.9
1988-89	909	5.8	17.1

It is statutory for the Boards to make a minimum profit; that is to manage by itself and be self sufficient. But now the return on their capital has only been in the negative. So poor is now its management that, internal resources[2] go dwindling and the Debt-Service Ratio(DSR)[3] of the boards lie much below a unit.

Conclusion

Complete restructuring of all power Management Units (PMU) is thus called for, for the optimum use of available resources. In the present book, the select energy experts, not only jot on the possible use of new energy sources for the future, they also correctly emphasise on a mass energy consciousness suggesting energy audits and other economical uses.

Decentralisation of power supply, substitution of efficient equipments, modernisation of industrial processes (for a reduction in their power consumption), regular energy audit and proper maintainance of motors/ machines (to reduce friction) can fetch us better power-tommorow.

Notes and Reference

1. *Rate of Return on Capital:* After covering depriciation and interest cost, all SEBs are supposed to earn a return of not less than 3 percent by the way of their net Surplus. It is a statutory provision and has been effected from April 1985. Still almost all SEBs show now negative rate of return, (resportedly because of the compulsion for subsidised power supply to some sectors).

2. *Internal Resources (IR):* It is the retained revenue surplus or deficit (i.e negative balance) plus depriciation and miscellaneous capital reciepts minus the repayment of loans.

IR of all Boards for financing their plans have been assessed, at the current rates (inclusive of Govt. subsidies), to be as follows;

Year:	1985-86	1986-87	1987-88	1988-89	1989-90	1990-91
IR (Rs.Cr.)	-25	+33	-387	-608	-1538	-1654

3. *Debt-Service Ratio (DSR)*: it is obtained by dividing gross operating surplus (or deficit) by interest and total loan repayment-liabilities. It indicates the extent to which debt servicing liabilities are covered by gross operating surplus.Subsidies are not considered while calculating DSR. When it equals or exceeds one unit, DSR then indicates the sign of a healthy state.

As per the latest report of planning commission, overall position for Boards is now that only 22 percent of the liabilities are being covered by the operating surplus (i.e. DSR=.22, for the year 1989-90).

4. **Reference:** Planning Commission report on he working of SEBs and EDs, September 1990, Govt. of India.

Contents

CHAPTER 1

Scope For Energy Conservation And Environment Management In India

R.N. Subudhi, Faculty of Management Studies, Berhampur University, Orissa-760007

Total power generated in India today is not sufficient to fuel satisfactorily the developmental process. Per-Capita power consumption for India is almost a tenth of that of the developed countries and half of that of the neighbouring China. While an all out effort is badly required to increase rapidly the total power generation, at the same time, we can not just afford to close our eyes on the negative aspects like environmental pollution, life hazards and other allied possibilities.

The present chapter suggests better environment management through energy conservation awareness. It presents statistical profile and also explores various scopes for energy conservation.

Introduction

The whole developmental process of a country is highly dependent on the unrelent supply of energy. With the growing pace and magnitude of the development process, the demand for energy also grows manifold. For most of the countries in the world today, energy scene has been quite adverse. Nature has not been impartial in providing the major primary sources of energy in equal or proportionate amount to all countries. While some countries move ahead through the optimum exploits of their most advanced technologies, some other have the envying position of having an abundant reserve of oil or other minerals. And as such, a Gulf-war or two would not really bother them. But the problem deteriorates for the rest large chunk of countries. For them, we find an ever widening gap between the demand and actual supply of energy. For India, where we do not have sufficient reserve of primary energy-sources, like oil, the energy scenario at present is totally dismal. At one point of time we were thinking of rationing our consumption of petroleum products. In the power sector too, we still have restricted supply.* We remember now with disbelief, the father of electricity, Edison, saying "I will make electricity to cheap that only rich can afford a candle". Neither, in reality now, it is cheap nor is dependable. While every effort is to be made to improve our power generation capabilities, at the same time, we have also to consider the possible negative impacts of present power technologies and consumption systems on our environment.

Dismal Energy Scene

Supply of primary sources of commercial energy in India is not sufficient. We have consumed 53.6 million tonnes of oil in 1990-91. Out of this, net import is 18.6 million tonnes which is 34.7 per cent of the total supply. This is of course a better figure when compared with those of earlier years. In 1970-71, out of the total supply of 18.5 million tonnes of oil, net import was 11.68 million tonnes (63.14 per cent). For coal also, where India is believed to have a vast reserve, we are not self-sufficient. Our supply of rich, processed coal is much below the total requirement. Out of the total supply of 226.1 million tonnes of coal (or 110.8 million tonnes of oil equivalent) in 1990-91, 5.1 million tonnes of coal was imported. Exhibit-I shows the growth of total supply of primary sources of commercial energy (coal, oil, natural gas, hydro and nuclear power) over the years. This supply between 1970-71 and

---

*See appendix-1, for the region-wise figures of deficits between power requirement and actual availability.

1990-91 increased by an annual average rate of growth of 5.77 per cent. Total energy supply, taking all the sectors into consideration, oil equivalent terms, is 187.83 million tonnes for 1990-91.

However, considering the vast population base of our country, total energy supply is not sufficient. The per capita energy consumption, when expressed in Kilograms of oil equivalent, is a little over 200 units for India. But the corresponding figures for Canada, USA, Australia and U.K. are 9156, 7266, 4821 and 3805 units respectively (all 1987 figures). Even in the neighbouring country China, where the population problem is equally worse, per capita energy consumption/supply is over two-and-a half times more than that of India. Appendix-2 gives an account of the per capita electricity consumption for different states.

Insufficient Power Systems

From power generation (or energy supply) to energy utilisation at the user's end, the whole power system now is not efficient enough to get us the optimum output from the available limited energy supply. On the capacity utilisation front of thermal power stations, the all India plant Load Factor (PLF) is now just around 56%. For developed countries PLF is now over 70%. As the reasons accounting for low capacity utilisation, Government, in their report (of Planning Commission, Power & Energy Division) on the working of SEB's and EDs, points out factors like lower plant availabilities, inadequate and poor-quality supply of fuels like coal, deficiencies in the generating equipments and lack of adequate hydel back-up.

Transmission and Distribution (T&D) losses are also very high for India. Table-1 gives the state wise figures of T&D losses for the last three years. The all India average T&D losses is now around 21 per cent. In developed countries such loss is within 10 per cent. Vast T&D network for the widely dispersed localities in difficult geographical areas, abrupt theft or illegal use of power, inefficient equipments are the main reasons for such a high loss.

Table 1
Trasmission And Distribution Losses (mkwh)

S.No	*Agency*	*1988-89*	*1989-90*	*1990-91*
(1)	*(2)*	*(3)*	*(4)*	*(5)*
State Electricity Boards				
1	Andhra Pradesh	3090	3380	3789
2	Assam	379	394	418
3	Bihar	1557	1555	1679
4	Haryana	1836	1611	1827
5	Himachal Pradesh	286	322	375
6	J & K	853	1281	1219
7	Karnataka	2472	2664	2768
8	Kerala	1379	1397	1435
9	Madhya Pradesh	3505	3059	3403
10	Maharastra	4657	4402	4800
11	Meghalaya	40	34	47
12	Orissa	1389	1282	1336
13	Panjab	2238	2707	3084
14	Rajasthan	2241	2316	2554
15	Tamil Nadu	3123	3333	3687
16	Uttar Pradesh	5785	5955	6728
17	West Bengal	1412	1480	1499
18	Gujrat	3382	3418	3792
	Total (SEBS)	39624	40591	44440
	Percentage of Loss (All Board Average)	(21.0)	(20.0)	(19.7)
Electricity Departments (EDs)				
1	Arunachal Pradesh	13	16	23
2	Goa	139	147	155
3	Manipur	34	35	37
4	Mizoram	13	14	15
5	Negaland	16	23	25
6	Sikkim	08	08	07
7	Tripura	43	55	63
8	DESU	1754	1770	1680
9	Pondichery	86	95	112
	Total (EDS)	2106	2163	2117
	Percentage of Loss (All EDs Average)	(25.5)	(23.8)	(21.2)
	Grand Total	41730	42754	46557

Even at the users end, hardly any care is given for the proper utilisation of energy. Exhibit-2, as supported by Table-2, presents the sector wise changing distribution of power generation and consumption between 1970-71 and 1989-90. Consumption in industrial sector has come from 67.6 per cent to 46.3 per cent of the total energy supply while, for agriculture sector, the share has gone upto 25 per cent from 10.2 per cent. As the agriculture consumption has been the trouble point for the power industry (because of the provision of providing power at highly subsidised rate) the increase in the share of agriculture sector has just magnified the loss. Estimated loss for the agricultural sector is 3801 crores of rupees for 1990-91. Figures for preceding two years were 3317 and 2733 crore rupees respectively. Again because of the use of crude technology and equipments, power consumption requirement is much higher than the actual need. When we account the total loss, starting from generation to actual utilisation at the user end, we only make use of a little of the original supply of energy.

Mis-Use in Agricultural Sector

It is estimated that (6), "100 tonnes of coal mined for the purpose of power generation provide only the energy equivalent of 5 tonnes of coal in the form of actual irrigation, when we use electric pump sets in the agricultural sector. A full 95 tonnes of coal is lost, in energy equivalent terms, in the process of power generation, transmission and distribution (T&D) and for our inefficient pump set and faulty irrigation system". Power supply to agricultural sector, like rural electrification programme, has been very costly for us, from utilisation point of view. Exhibit-3 gives the statewise as well as all India figure of energised pump sets/tube well. As evidenced from Table-(2) and Exhibit-2 gradually more and more percentage of energy are being consumed in agriculture sector. Exhibit-3 shows the very high growth of energised pump sets. Populist, low electricity tariff for agricultural consumptions has virtually led to gross misuse of power. In the agricultural sector alone, with due care, a large amount of power could be avoided/saved. From the Karnataka experience (Ref. R. Sriram

Table 2
Trends in Sector-wise Power Consumption (Percent)
(Utilities only; 1960-61 to 1990-91)

Year	*Domestic*	*Commercial*	*Industrial*	*Rly/Tram*	*Agrl*	*Others*
1950-51	12.6	7.5	62.6	7.4	3.9	6.0
1960-61	10.7	6.1	69.4	3.3	6.0	4.5
1970-71	8.8	5.9	67.6	3.2	10.2	4.3
1975-76	9.7	5.8	62.4	3.1	14.5	4.5
1976-77	9.5	6.2	62.5	3.3	14.4	4.1
1977-78	9.9	6.4	61.6	3.3	14.6	4.2
1978-79	9.8	5.6	61.8	2.8	15.6	4.4
1979-80	10.8	6.0	58.9	2.9	17.2	4.2
1980-81	11.2	5.7	58.4	2.7	17.6	4.4
1981-82	11.6	5.8	58.8	2.8	16.8	4.2
1982-83	12.7	6.1	55.4	2.8	18.6	4.4
1983-84	12.9	6.4	55.8	2.6	17.8	4.5
1984-85	13.6	6.1	55.2	2.5	18.4	4.2
1985-86	14.0	5.9	54.5	2.5	19.1	4.0
1986-87	14.2	5.9	52.5	2.4	20.7	4.3
1987-88	14.7	6.0	48.8	2.5	23.8	4.2
1988-89	15.5	6.2	47.1	2.3	24.3	4.6
1989-90	16.3	5.8	45.6	2.4	25.0	4.4
1990-91*	16.5	6.1	44.9	2.2	26.0	4.3

*Provisional

Source: 1. Economic Survey 1991-92, Govt. of India.
2. Current Energy Scene in India (July 1990) CMIE Bombay.

Exhibit-1

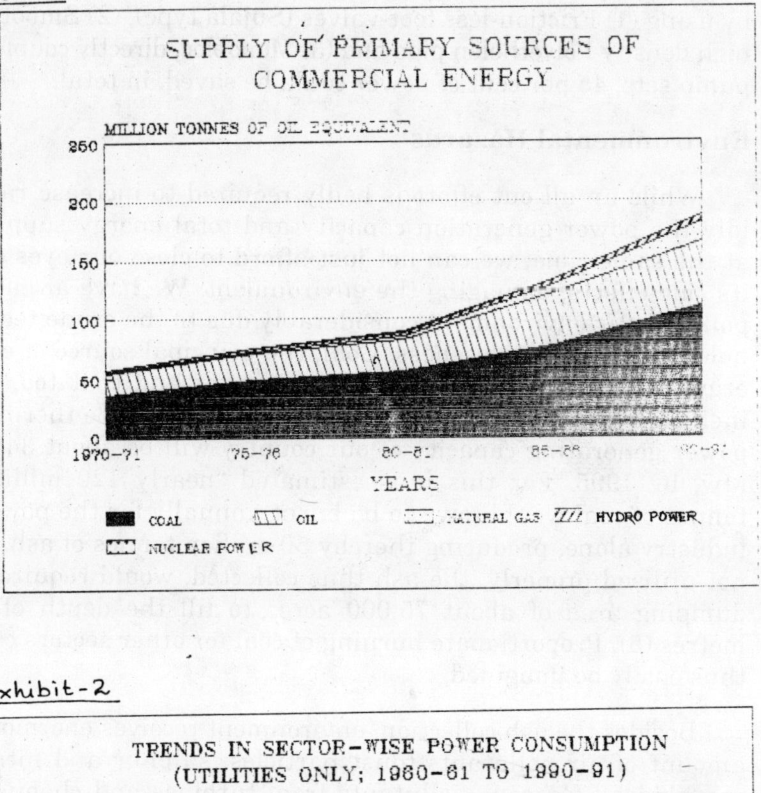

Exhibit-2

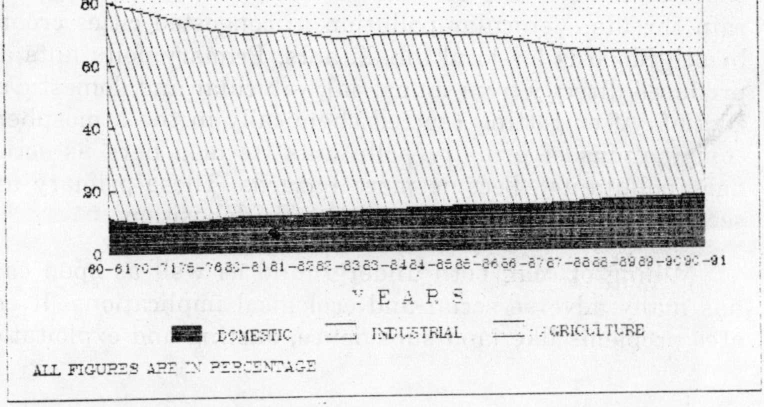

in URJA, April 1992, Page-25), it has already been found that by using (1) Friction-less foot-valves (Sujala type), (2) Smooth, high density Polyethelen pipe and (3) Monobloc, directly coupled pump sets, 45 per cent of power could be saved, in total.

Environmental Hazards

While an all out effort is badly required to increase rapidly the power generation capacity and total energy supply, at the same time, we can not just afford to close our eyes on its side-effects, damaging the environment. We have already polluted our environment considerably due to the crude technology and our carelessness. Coal, the principal source of energy, is being burnt daily in huge amount and that too, in inefficient manner. According to a CEA estimate, the thermal power generating capacity of our country will be about 3400 Mw, by 1995. For this it is estimated "nearly 120 million tonnes of coal would have to be burnt annually for the power industry alone, producing thereby 50 million tonnes of ash. If not utilised properly, the ash thus collected, would require a dumping area of about 75,000 acres to fill the depth of 5 metres (5). Proportionate burning of coal for other sectors can thus easily be imagined.

Besides the ash collection, environment receives enormous amount of air pollutants (dust particles, sulphur and nitrogen oxides). Gaseous pollutants from turbines and chemical industries are more dangerous. We have been fortunate, till today, by not having acid rain problems. Exhibit-4 gives the chart of some of the contents of air pollutants, posing acid rain threats. Vehicular pollution in congested cities creates health hazards for local inhabitants. Primary pollutants are produced directly from industrial, vehicular and domestic activities. After getting reacted chemically in the atmosphere, combinations of primary pollutants (termed here as secondary pollutants), become more hazards. These primary and secondary pollutants comprise the acid deposition.

Mining of coal, both underground as well as open cast, has many adverse social and ecological implications. It creates problems like land subsidence, ousting and exploitation

Exhibit - 3

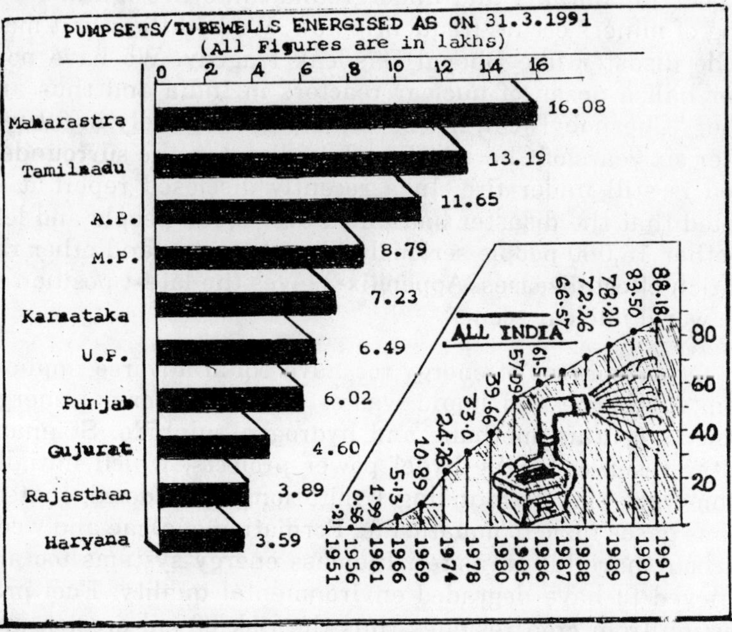

State-wise figures for other states, which are below 3 Lakhs, are:
BIHAR (2.55), KERALA (2.22), WB (0.89), ORISSA (0.51), GOA (0.04),
ASSAM (0.04), HP (0.03), J&K (0.02), TRIPURA (0.01) and UT's(.32).

Exhibit - 4:

ACID RAIN THREAT FROM AIR POLLUTANTS

	PRIMARY	→ SECONDARY
GASEOUS POLLUTANTS	Sulphur dioxide(SO_2) Nitric oxide(NO) Nitrogen dioxide(NO_2) Hydrocarbons Ammonia(NH_3), CO_2	→ NO_2 from oxidation of NO Ozone(O_3) and other photo-chemical oxidants formed by action of sunlight in mixtures of NOx and hydrocarbons.
PARTICULATE POLLUTANTS	Fuel ash Metallic particles	→ Sulphuric (H_2SO_4) and nitric (HNO_3) acids formed by oxidation of SO_2 and NO_x. Reaction products (NH_4SO_4, NH_4NO_3 etc) of sulphuric and nitric acid with NH_3.

of poor inhabitants and underground mine fire. Fatal trapping of miners occurs frequently now. We had the worst man made disaster like Bhopal Gas-leak tragedy. We have now over half a dozen of nuclear reactors in India and thus another "Chernobyl" can not be ruled out completely in future. After six years of Chernobyl nuclear disaster, the surrounded area is still under fire. In a recently disclosed report it is stated that the disaster had killed over 5,000 people and left another 15,000 people seriously ill with cancer and other radiation-effect diseases. Appendix-3 gives the latest position of the world nuclear scene.

Other sources of energy too have equal adverse implication. Air borne and liquid wastes from geo-thermal energy projects contain methane and hydrogen sulphide. Stagnant water reserves of the hydel power projects, which initially submerges the surrounding habitations, gets deoxigenated, forms gases toxic to aquatic life. Formation of algae and weed harbour harmful parasites. Biomass energy systems too are believed to have degraded environmental quality. Fuel production from crop residues runs the risk of soil erosion and water run-off. That is why now every power project requires environmental clearance from the concerned ministry/department.

Scope For Conservation

Besides new-energy additions, we have now two more things to look after, which seem to be very much related. First, to check the heavy loss of energy through improved technology and a combined, spirited effort and second, reducing substancially the environmental threats of power projects and modern machineries. That is to conserve energy and to protect ourselves and our surroundings.

On conservation front, participation of both public as well as Government/administration is essential. Proper and regular maintenance of vehicles would not only give us extra mileage (and save oil), it would also keep the environment of our cities pollution free. Saving of a drop of oil per Vehicle per day would result in a national saving of thousands of litres of oil. By using high grade lubricant and friction-less foot-valves in irrigation pumpsets, 45 per cent of energy can be saved. As shown in exhibit-3, we have 88.18 lakhs of energised pumpsets (as on 31.3.1991). Thus a massive amount of energy can be saved in this sector. Many research works have been carried out on the conservation scopes in each sector. Reddy and others (9) have recently, after identifying six types of consumers, prescribed conservation possibilities for each case. Replacing inefficient incandescent bulbs by compact fluorescent lamp (CFL) or tubes (which consumes only one-fourth of energy as consumed by the former), great amount of total energy can be saved. This gives better light too. While planning a building, care should be given for the optimum design of power points too. Schemes like improved smokeless Chullah, biogas plants, solar heaters (for supplying hot water) should draw more Government support and importance.

By removing extraneous substances near the mines, improving the efficiency of washeries and modernising the industrial processes and (thermal) power generation units, fly ash can be reduced considerably. Fly ash can also be utilized for different purposes like making bricks, land/mine filling, construction and other chemical uses (5). Through modernisation, Rourkela Steel Plant (RSP) has brought down its energy consumption rate per tonnes of crude steel from a level of 12.76 gega calories (gc) in 1983-84 to 10.06 gc in 1990 and thus saved over Rs. 200 crore. Consumption of coke per tonne of hot metal has also been reduced from 886 Kg in 1983-84 to 728 Kg in 1991-92. However only a few industries have so far felt for the need of energy and its energy efficiency measures. Works (11) rather reveal a higher energy dependence of many industrial processes. (Appendix-4, gives the statistical account of increasing energy dependence of industries.)

The colossal T&D loss, as listed in Table-I, can be checked by 50 per cent. The rate of T&D loss, which is now for us over 21 per cent, is infact just around 10 per cent for the countries with advanced technology/equipments. However, this is not a simple affair, rather involves very high cost. For this, as estimated by the department of power, Rs. 5000 crore would be required. Decentralisation of power supply, drawing of more EHT line, efficient T&D equipments, together, can check the loss.

Conclusion

Energy saved is as good as energy generated. A kilowatt hour saved is just not equal only to a kwh generated; but much more than that, as we consider factors like T&D losses. At the current rate of T&D loss of 22.5%, one Kwh saved at consumer end is thus equal to 1.29 Kwh generated at the generation end (9). If by replacing all filament electric bulbs by energy efficient (compact) fluorescent tubes, a huge amount of power could be saved, equalling the total nuclear power production of India, (and thus averting any nuclear disaster) it is possibly the best way of protecting our environment and at the same time meeting our all energy demand, by conserving energy. What is called for urgently now is a mass awareness on the conservation need a plan for complusory energy education. Common-man's participation in energy conservation programme and practical execution is thus much more important.

REFERENCE

1. DK Bose, *Planning for Power Development in Orissa and W.B.*, Urja Vol. 31 (3), March 1992.

2. RC Das & LN Patnaik, *Energy Scenario in the State of Orissa: A Status Report,* Energy Environment Monitor (TERI) Vol. 7 (4), March 1991.

3. Richard Eden et al, *Energy Economics* (Ch. 13, pp 311-329), Cambridge 1981.

4. A Ghose, *Eigth Plan: Challenges and Opportunities -VI (Energy Conservation and Demand Management),* Eco 2 Pol. Weekly Feb. 23, 1991.

5. Mukesh Kumar, *Recycling of sewage and fly-ash on land,* Ph.D. thesis submitted to JNU New Delhi, 1991-92.

6. RK Pachuri, *Population and Energy,* in Gowariker's (Ed), *The inevitable billion plus: Science, Population And Development,* Unmesh Com., 1992.

7. Pagedar & Singh, *Environmental Protection by Retrofitting Correctly Specified Electrostatic Precipatators,* Urja Vol. 31 (1), Jan, 1992.

8. TK Ray, *Air Pollution Control Systems in Thermal Power Plants,* Urja Vol. 31 (3), March, 1992.

9. AKN Reddy et al, *A Development-focused End-use-oriented Electricity Scenario for Karnataka,* Eco. & Pol. Weekly, Vol XXVI (14 & 15), April' 91.

10. RK Sharma, *Atmospheric Pollution Control Scenario of TPS in India: Formulation and Promulgation of Environmetal Standards,* Urja, Mar'92.

11. RN Subudhi, *Changing Energy Scenario of India: A Review,* Urja Feb'92.

Appendix - 1
Demand And Supply Of Power (Million Units)

	NR	*WR*	*SR*	*ER*	*NER*	*ALL INDIA*
1984-85						
Requirement	44722	46430	39729	22957	1594	155432
Availability	38471	46118	39810	18982	1632	145013
Deficit(-), Surplus(+)	-6251	-312	+81	-3975	+38	-10419
% of Deficit, Surplus	-14.0	-0.7	+0.2	-17.3	+2.4	-6.7
1985-86						
Requirement	48504	51311	45535	23643	1753	170746
Availability	43303	50595	41298	20376	1690	157262
Deficit(-),Surplus(+)	-5201	-716	-4257	-3267	-63	-13484
% of Deficit(-)	-10.7	-1.4	-9.3	-13.8	-3.6	-7.9
1986 - 87						
Requirement	55006	57694	51178	26549	1929	192356
Availability	49807	55263	45536	21860	1810	174276
Deficit(-), Surplus(+)	-5199	-2431	-5642	-4689	-119	-18080
% of Deficit(-)	-9.5	-4.2	-11.0	-17.7	-6.2	-9.4
1987 - 88						
Requirement	62891	63135	55286	27560	2121	210993
Availability	55801	60024	45927	24205	2019	187976

(Contd.)

Contd from Appendix - 1

	NR	*WR*	*SR*	*ER*	*NER*	*ALL INDIA*
Deficit(-), Surplus(+)	-7090	-3111	-9359	-3355	-102	-23017
% of Deficit(-)	-11.3	-4.9	-16.9	-12.2	-4.8	-10.9
1988 - 89*						
Requirement	59759	60335	54231	26135	2251	202711
Availability	56179	58754	46819	23112	2174	187038
Deficit(-), Surplus(+)	-3580	-1581	-7412	-3023	-77	-15673
% of Deficit(-)	-6.0	-2.6	-13.7	-11.6	-3.4	-7.7
1989 - 90*						
Requirement	68219	67915	59343	28018	2519	226014
Availability	64213	66184	51429	23815	2443	208084
Deficit(-), Surplus(+)	-4006	-1731	-7914	-4203	-76	-17930
% of Deficit(-)	-5.9	-2.5	-13.3	-15.0	-3.0	-7.9

*Accounted Period: April-February.

NR = Northern Region, comprising Haryana, HP,J&K, Punjab, Rajasthan, UP, Delhi, and Chandigarh.
WR = Western Region, comprising Gujarat, MP, Maharastra, Goa.
SR = Southern Region, comprising AP, Karnataka, Kerala, TN and Pondichery
ER = Eastern Region, comprising Bihar, WB, Orissa and DVC.
NER = North-Eastern Region.

Source: Current Energy Scene in India, July 1990, CMIE Bombay.

Appendix - 2
Per-Capita Consumption of Electricity
(Utilities & Non-utilities, units in KWH)

States / UT's	*1970-71*	*1980-81*	*1987-88*	*Annual growth (%)*
Punjab	159	315	515	7.2
Goa (Daman, Diu)	92	233	357	8.3
Maharastra	158	272	347	4.7
Gujarat	138	245	373	6.0
Haryana	98	213	306	6.9
Tamil Nadu	130	190	249	3.9
Karnataka	104	157	207	4.1
Andhra Pradesh	56	106	197	7.7
Madhya Pradesh	54	98	187	7.8
Orissa	96	116	165	3.2
Jammu & Kashmir	38	73	163	8.9
Himachal Pradesh	35	66	162	9.4
Rajasthan	50	99	162	7.2
West Bengal	118	114	135	0.8
Utter Pradesh	60	87	135	4.9
Kerala	76	101	130	3.2
Bihar	65	76	104	2.8
Assam	22	34	64	6.4
Meghalaya	-	31	88	16.1
Sikkim	-	43	65	6.1
Negaland	8	32	62	12.8
Manipur	5	6	60	15.7
Arunachal Pradesh	-	15	47	17.7
Mizoram	-	7	41	28.7
Tripura	5	13	35	12.1
Union Territories:				
Delhi	286	402	567	4.1
Chandigarh	287	307	483	3.1
Pondicherry	180	270	439	5.4
Dadra & NH	9	64	420	25.4
Lakshadweep	7	27	137	19.1
Andaman	26	42	112	9.0
All India	90	135	201	4.8

Source: Basic Statistics of Indian Economy, Sept'1990, CMIE Bombay.

Appendix - 3
International Nuclear Power Position
(Till the end of 1991)

Countries	*No. of Nuclear Plants*		*Installed*	*Electricity Supplied*		
	In Operation	*Under Construction*	*Capacity MW(e)*	*Nuclear Power in'91*		
				MW(e)	*Twh*	*% Total*
USA	111	3	99757	3480	612.6	21.7
France	56	5	56873	7005	314.9	72.7
USSR (erstwhile)	45	25	34673	21255	212.1	12.6
Japan	42	10	32044	9192	209.5	23.8
Germany	21	—	22390	—	140.0	27.6
Canada	20	2	13993	1762	80.1	16.4
U.K.	37	1	11710	1188	62.0	20.6
Sweden	12	—	9817	—	73.5	51.6
Koear Rep.	9	3	7220	2550	53.5	47.5
Spain	9	—	7067	—	53.2	35.9
Belgium	7	—	5484	—	40.4	59.3
S.Arfica	9	—	1842	—	9.1	5.9
India	7	7	1374	1540	4.7	1.8
Argentina	2	1	935	692	7.2	19.1
Mexico	1	1	654	654	4.1	3.6
Brazil	1	1	626	1254	1.3	0.6
Pakistan	1	—	125	—	0.4	0.8

Source: Urja, Vol.31(5), May 1992.

Appendix - 4
Fuel Consumption Scene for Manufacturing Sector

	Major Consumers	*1975 - 76* No. of Factories	Consumption Rs.Lakh	Dep* (%)	*1985 - 86* No.of Fact.	Cons. Rs.Lakh	Dep* (%)
1	Iron & Steel	1150	24725	13.4	1764	119420	13
2	Cement, Lime, Plaster	205	9777	27.3	468	63652	26.7
3	Spinning & Weaving	1018	15988	6.3	1349	58751	10.1
4	Fertiliser, Pesticide	378	9137	10.2	546	57955	14.7
5	Organic & Inorganic	639	6859	13.7	1040	53112	19.9
6	Pulp, Paper board	424	5215	12.1	802	32623	19.4
7	Aluminium	200	4131	18.1	331	30100	33.5
8	Iron, Steel Casting	2737	6096	7.7	3337	26763	8.3
9	Fibre, Plastic, Synth.	167	3909	9.4	243	16658	9.6
10	Strl. Clay Products	1533	2622	21.8	3421	12629	23.6
11	Glass-wares	414	2087	17.7	571	11296	23.7
12	Ferro alloys	42	1333	28.1	54	8007	31.7
13	Porcelain, China-ware	291	789	19.7	567	4077	26.1
14	Copper	111	771	8.3	144	3006	10.7
15	Zinc	11	212	5.9	46	2885	14.5
16	Starch	183	284	7.5	284	1593	13.4
17	Ice	242	191	33.3	233	720	32.4

(Contd.)

Contd. from Appendix - 4

Major Consumers		1975 - 76 No. of Factori	Consumption Rs.Lakh	Dep* (%)	1985 - 86 No.of Fact.	Cons. Rs.Lakh	Dep* (%)
18	Common salt	282	75	6.6	414	339	7.9
19	Slaughttering, Meat p	13	21	4	32	291	7.7
20	Dyeing, Bleaching Tex	69	92	10.9	83	198	15.7
21	Sub Total	10109	94314	--	15729	504075	---
22	Other Sectors (org)	58150	47592	3.7	81179	583634	8.70
All Factories(Total)		68259	141906	5.1	96908	1087709	10

*Dependence on fuel= (Total Fuel Consumed/Value of Output)*100
Annual Growth Rates:(1) No. of Fact.=3.6,(2) Fuel Consumption=22.6
Source: "Current Energy Scene in India", July 1990, CMIE Bombay

CHAPTER 2

Energy Policy and Planning for Future : Some Essential Considerations*

Prof. R. Natarajan
I.I.T. Madras.

This chapter discusses several important issues in the areas of energy policy and planning for the future, with particular reference to developing countries like India. The correlations between energy use on the one hand and national income, standard of living and quality of life on the other, are examined. The special role of electricity in end use and the role of renewable energy sources in energy supply are discussed. The physical and social economy of energy use and the ethical and moral aspects of energy use are examined. Some future trends, and the expectations from technology are delineated. Some of the special characteristics of strategies for developing countries are finally discussed, leading to the conclusion that developing countries need not retrace the stages of economic development in industrialized countries.

There Appears to Exist a Strong Historical correlation between energy use and national income for most countries. Using 1961 data on all the 153 countries existing at the time, Felix(1) showed that national income per capita and total commercial energy consumption per capita are closely related. More recently, Darmastadter et al (2) have looked at this re-

---

* Reproduced, with kind permission, from Urja, Vol. 27(1). January 1990.

lationship. Their basic conclusion is that the per capita consumption of commercial energy forms bears a quantitative relationship with indicators of general economic development, measured by per capita of general economic development, measured by per capita GNP, i.e. an area's production of all goods and services per person. This relationship between GNP and energy is considered to hold both cross-sectionally and historically.

However, the chain of causation between these two factors is less distinct. Energy has been called the fuel of economic progress, and the ready availability of cheap energy serves to stimulate industrial development. Conversely, progress in energy-intensive industrial development will lead to enhanced energy production and consumption per capita. The direction of causation can not be established by the existence of a statistical relationship between income and energy consumption. However, it is useful to examine the degree of correlation:

Employing income per capita (I/P) and commercial energy consumption per capita (E/P) for 1965 for 49 countries having per capita income of atleast \$ 300, Darmastadter obtained the following equation:

$$\log (E/P) = 1.12 \log (I/P) - 0.34 \quad ...(1)$$

Correlation coefficient = 0.87

He obtained the following equation for 24 countries with per capita income levels ranging between \$ 300 and \$ 899:

$$\log (E/P) = 1.79 \log(I/P) - 1.93 \quad ...(2)$$

correlation coefficient = 0.70

Such relationships have not been worked out for less developed countries (LDCs) with per capita income much below \$ 300.

Star (3) obtains the following historical GNP energy relation for the US:

$$E = 0.0807\ GNP + 5.495 \quad ...(3)$$

Tyner (4) has investigated this relationship between total commercial energy consumption and national income (at 1961-62 prices) for the 18 year period from 1953-54 to 1970-71. Both log-log and I linear forms gave very good fits. The linear form is:

$$I = 68.90 + 0.592\ E\ ...(4)$$
correlation coefficient = 0.99

Here, E is the total energy consumption in million tons of coal replacement, and is national income (at 1960-61 prices) in billions of Rupees.

The log-log form is:

$$\ln I = 2.630 + 0.592E \quad ...(5)$$
correlation coefficient = 0.99

Thus, it is clear that the correlation is widespread and is valid for several stages of development. As far as the casual relationship is concerned, it might be argued that increasing economic activity stimulates greater energy consumption. However, it must be recognized, particularly in the Indian context, that inadequate supplies of energy will inhibit development, and that assurance of an adequate supply and mix of energy inputs will be a great stimulus to development. Conversely, inadequate energy supplies will inhibit economic growth. We have, for example, seen how power shortages have affected a wide range of economic activity in both the industrial and agricultural sectors.

It is also important to recognize that about one half of the energy consumed in India is non-commercial energy, comprising firewood, dung and agricultural waste products, which are usually gathered outside the market economy. The share of non-commercial energy consumption has been steadily falling. This trend is expected to continue as development proceeds.

Goldemberg et al (5) indicate that far more important than the sheer magnitude of GDP growth are the structure and content of growth and the distribution of its benefits for achieving effective economic growth for the majority of people. They caution that the assumption that improving human welfare necessarily requires greatly increased energy use should not be blindly accepted. While it is simplistic to assume that energy use must increase with the level of economic activity, energy use depends also on the mix of energy-using activities in the economy and on the technologies used to provide energy services. In turn, the mix of energy-using activities depends on the development path chosen, and the energy requirements of a particular energy service depend on the energy efficiency of the end-use technology selected.

They point out that the world has already witnessed a decoupling of energy and economic growth. Between 1973 and 1985, the total energy use per capita in the OECD countries fell 6% while the per capita CDP increased 21%. In the US, per capita energy use fell 12 %, while per capita GDP increased 17%: the corresponding figures for Japan were 6 and 46%.

Such decoupling is expected to continue into the future, both because of ongoing structural shifts in their economies and because of opportunities for more efficient energy use.

It is also pointed out that for oil-importing countries, the exchange rates used to convert their national currencies into US dollars play a significant role, in this context. These exchange rates do not often reflect the purchasing power of the currencies of the developing countries at home. For example, the equivalent a dollar can buy a lot more in a developed country than it can in the US.

There are several factors which can distort the GNP-energy relationships. The energy requirements associated with environmental protection would increase the demand for energy. For example, the following steps result in increased energy use and/or decreased conversion efficiency:

(i) Changing from water-cooled condensers, to reduce the detrimental effects on water resources.

(ii) removal of undesirable chemical from coal prior to combustion.

(iii) restrictions on emission of sulphur, in the non-electric sector.

On the other hand, several technological innovations have caused the energy input per unit of GNP to drop significantly. Some examples of technology substitution are given below:

	Substitute	Replacement for
i	electric motor drive into manufacturing	prime mover-belt-drive
ii	diesel locomotive	steam locomotive
iii	fluorescent light	filament lamp
iv	transistor	vacuum tube

Indicators of Economic Progress and Development

Stewart (6) consider four indicators as valid indices of economic prosperity:

i gross national product (GNP)
ii gross national energy consumption (GNEC)
iii employment (or unemployment) statistics
iv price indices of consumer products.

Among these, it is not very obvious that prices are directly related to GNP or economic well-being. Also, prices can be strongly affected by other social disorders, such as wars, cartels, boycotts, and pestilence. The two long periods of prosperity during this century in the US, viz., during 1900—1930 and 1955—1985, and the one period of severe depression from 1930 to 1940, bear correlation with GNP, GNEC and unemployment statistics. Among those, the size and intensity of the work force, revealed by the employment statistics, do not take into consideration the manufacturing contribution of machines.

That leaves GNP and GNEC. The GNP is a measure of all the nation's end products being marketed, while business growth has a direct and profound effect on one of society's most fundamental resources, viz., the total energy ued by the nation. While both have certain weaknesses as absolute indicators of societal well-being, it is important to note that a strong correlation exists between the two. The GNEC can be estimated reasonably accurately by adding the energy input contributions from all the fuel resourcs-firewood, coal, oil, natural gas, nuclear fission, hydroelectric, wind, solar energy, etc. As the living standards have improved, per capita energy use has gradually increased. The overall national energy consumption. GNEC has also increased over the year, partly because of the per capita energy increase, and also because of the large population increase.

In conventional economics, while GNP a is popular indicator of well-being, GNEC growth is considered to be a key indicator of national vitality. The quantity and quality of energy consumed by a society is considered to provide a better indication of its prosperity than GNP, both for the same country from year to year, and from country to country. The decline of GNEC since 1979 in the US has been attributed mainly to a successful national energy conservation effort, and partly to higher energy prices.

Stewart (6) has elaborated on the roles played by GNP and GNEC, especially their growths. The degree of economic prosperity can be defined by the total commerce of goods and products within the society. In a consumption-oriented society, it is customary to equate well-being with national economic wealth, as measured by GNP. The years in which GNP has shown a growth of 4 to 6 per cent have been haralded as years of prosperity in the US, GNP, measured in inflation—adjusted dollars, is expected to continue to be the traditional index for measuring the dollar value of goods and services. It can be measured in one two ways:

(i) by the final market value of output of goods and services.

(ii) by the total income accruing to salaried employees, business, owners, etc.

"Societal industry" is defined as the total applied industry within the society associated with all human activities related to goods and services, including those for commerce, government and the home. Three alternative measures are suggested for quantifying it:

(i) the monetary value of all goods and services produced, as measured by an index like GNP.

(ii) the total employment level in the workforce, as measured by man-hours of work.

(iii) the total amount of energy consumed by society to sustain itself.

Each of the above has certain merits and shortcomings. The first measure may not be accurate, since the prices of end products and services, as well as the incomes of individuals, are affected by supply demands considerations. The second measure ignores the industrial values of different work enterprises and overlooks the contribution from capital equipment used for the mechanization of work, such as machine tools, robotics, computer technology, etc. This leaves the total energy embodied in the goods and services as the most valid measure which can be assigned to overall societal industry. The societal energy use includes both the direct energy expended for various activities and the embodied energy in all the tools devised and used by the society.

The monetary value of energy resources accounts for less than 10 per cent of GNP in the US: this leads to a gross under-estimation of the full importance of energy. The money value associated with energy resources does not even represet an inherent value of energy. The money value simply represents the costs of labour, capital equipment and energy itself, to locate, extract, process, and distribute the energy resources to suitable and points for purchase and use by society. In so far as industry (in the broadest sense) depends on the availability and use of energy, GNEC appears to be the most appropriate measure of societal industry.

Starr (3) points out that among the societal targets which are most directly affected by energy relationships are health, security, social stability, upward mobility and leisure. The traditional target of "economic growth" may be one of the best composite means of achieving all these materialistic goals, and this explains its frequent and popular use as a measure of societal success. Other contributors to social development have been greater productivity in agriculture, improved transportation and communications, and shorter working hours.

Another approach has been to develop a composite index for technology, comprising power, steel, transportation engineers and scientists, and communications to represent the technological status. A separate socio-economic composite index has been defined to include such values as GNP per capita, university students, infant mortality, mental health, unemployment, urban crime, etc., and is shown to correlate with the technological index.

Goldemberg et al (5) relate development to:

(i) the satisfaction of basic human needs
(ii) industrial activities that promote employment
(iii) agricultural production.

The standard of living is related to food, shelter, sanitation, health, education, and gainful employment.

They point out that while growth is not to be considered an end in itself, but rather a performance test of development, development tends to be equated with growth, and the goal of development is considered to be the maximixation of the volume of goods and services produced by the economy.

There is another approach to development, the basic needs approach, in which some of society's scarce resources are directed toward the satisfaction of a well-defined set of the most basic human needs. Among economists and planners, the definition and quantification of basic human needs have long been debated. There is consensus, however, on the need to ensure minimum standards of nutrition, shelter, clothing, sanitation, health, education, and gainful employment.

Correlation between Energy use and Standard of Living

Energy system studies have revealed that out quality of life, material welfare, health, employment, and income are all demonstrably affected by energy availability and cost. In a developed industrial society, there exists a set of inter-relationships that tie together energy use and a wide range of physical, economic, and social aspects of life.

Two questions are often asked in connection with this correlation:

(i) Is a high GNP per capita really necessary for people's welfare and happiness?

(ii) Is the use of energy in achieving a high GNP per capita as efficient as it should be?

It is interesting to observe, in this, context, that the education, creative, and leisure fraction of a man's lifetime in an industrial society is estimated to be about 45% as compared with 25% for agricultural man, and 17% for primitive man. The average life expectancy of an industrial man is twice that of agricultural man, which in turn is almost twice that of primitive man. It is postulated that there is a relationship of these factors to the goods and services available in each society.

Starr (3) has investigated the interpretation of the meaning and significance of GDP/E ratios, and has examined whether these comparisons measure the relative efficiency of energy use. While considering the relationship of these ratios to life-styles, the area of concern is the consumption side of GDP, and not the generation of product. The way we live, or the "habits" of consumption, assume importance when the conservation of energy becomes an issue. Based on a composite index of social, economic, political and environmental variables, California ranked number 1, and Alabama number 50, in an inter-state comparison in the US. On an income per capita basis, California ranked number 5, and Albama number 44. Thus, per capita income may not serve as a sufficient measure to explain the "well-being" of people, and the overall environment in which they live.

In a similar study for different countries, a composite index was developed to include the number of telephones, cars, radios, and television sets in use; the share of expenditures for food, drink, clothing, rent, recreation and energy in personal consumption; etc. This index was expected to quantity lifestyles in industrialized countries. The study revealed wide differences in lifestyle among, for example, the EEC countries. It is also shown that changes in lifestyles are partially brought about through increased uses of energy: in other words, there is a positive relation between energy and the way people live.

While it is conceded that energy consumption will be reflected in any given society's lifestyle, lifestyles mean different things to different people, and at present no consensus exists as to what it is or what it means.

Cook(7) points out that it is the product of population and living level that is limited by the availability of energy, and that the impact of an industrial growth society on energy sources is geometric, because both population and living levels grow.

Correlation between Energy use and Quality of Life

It is often pointed out that we should distinguish between level of consumption and level of life: the former defined by access to cars. TVs etc., and the latter by some measure of the average enjoyment of life through the yardsticks of health, education and leisure. GNP is considered to be inadequate to represent all the values inherent in the meaning of "standard of living"

Philosophers like Rousseau have proclaimed that a primitive society, in "harmony with nature", uncluttered by the artifacts and unhampered by the strictures of civilization, has the highest standard or level of life. Anthropologists, however, have shown that most primitive peoples lived short lives of fear and discomfort, and that they appeared to live in harmony with nature only because their means for destroying nature were comparatively important. Since the days of Adam Smith, several persons have opposed the concept of the "noble savage", contending that continuous growth in the use of energy and materials is the best state for man.

While many people consider it a fallacy to link the quality of life causally with per capita access to energy and materials, this link has been the philosophical foundation of western industrial, technological, democratic society; and political decision makers act on the basis of its validity.

Starr (3) concedes that social values have proven extremely difficult to quantify, and that their interpretation is highly subjective, but a broad consensus is considered to exist in our culture for the following goals:

(i) meeting the basic material needs of all members of society.
(ii) equality of opportunity in all areas
(iii) a clean environment
(iv) freedom from illness and risk
(v) basic personal freedoms
(vi) availability of satisfying employment
(vii) security, both from wars and from crime
(viii) providing the needs of the elderly

consumption to the perpetuation of productivity, or supply. While in such an economy, it is easy to define waste and to set up a moral code that forbids waste, there is a wide range of opinions on the definition of waste in an economy of abundance.

Cook considers energy use, environmental sanity, and the quality of life as independent variables. Each variable is considered to have at least one real limit. The limit on energy use is a maximum, meausred as the rate of availability of useful energy that can be sustained during any chosen period. The limit on environmental sanity can also be considered to be a maximum, either of pathogenic pollution or of uncontrolled disease organisms, in either case determined by the ability of the human species to persist. The quality of life also has a maximum, namely, the number of persons that can be supported by the available food supply or that the social mechanism can tolerate before it destroys itself. The two most threatening social costs of present energy systems are environmental poisoning, and social collapse due to the exhaustion of nonrenewable resources.

The Special Role of Electricity as an Energy form

Electric power is one of the most important components of the energy sector. The correlation between power, consumption and the index of industrial production in India for the years 1955-56 through 1970-71 was 0.96 (4). Power production and consumption are also vital for the agricultural sector.

Electricity is an energy form, not a source. It can be generated from any of the raw energy sources, transmitted and distributed with relative ease, and converted into useful work, heat, or light at any end-use point. Although it can be stored in small quantities by electrochemical reconversions, it can not as yet be stored as electricity in any appreciable quantities. Apart from this (significant) disadvantage, it is the perfect energy intermediary. It is quite unsubstitutable for some specific end uses; for example. TV, telephone, fluorescent lighting, computers, etc.

Perhaps, one of the most important advantages of electricity is that it has the greatest potential for permitting efficient conversion of energy sources into end use functions. In space and water heating, direct use of fossil fuels and solar energy compete, but even here, because of its ease of control and flexible location, electricity is often preferred.

Stewart (6) shows that the fraction of energy used for electricity generation shows a stronger correlation with GNP than total energy. The ratio Eel/GNP for the US (Eel referring to the input energy to electricity production) shows a positive growth rate over the years, implying that electrical energy is growing faster than GNP. This means that a larger fraction of the GNEC is being used each year for electricity generation. Since GNEC/GNP shows a declining trend, this shows that the electrical energy is being used more effectively than the energy previously used for the same purpose.

Cook (7) argues that electrification is a result, not a cause, of national affluence. Electricity in the national energy consumption is a meausre of living standards, supported mainly by energy surpluses derived from the fossil fuels.

Goldemberg et al (5) fault the conventional wisdom among planners in developing countries who equate development with economic growth, and economic growth with growth in energy use, with emphasis on electricty, especially electricity produced in large central power plants.

Typer (4) points out that one means of conserving electric energy is by modifying the rate structure. Often, power rates decline with consumption, giving large discounts to high volume consumers. This type of rate structure provides an incentive to consume larger amounts of power, because the marginal cost declines with rising consumption.

The Physical Economy of Energy use

Physical economics is the study of the benefits and costs of man's activities in terms of the physical laws that govern change: such change invariably involves energy. A lesson of general application is that life is cyclic as far as materials are concerned, but is undirectional as regards energy. It is important to distinguish between different benefits and costs to man as a result of energy transfers and conversions. For example, sports and recreationary activities are energy dissipating, while agriculture is energy producing: the former represent an unproductive discretionary expenditure of accumulated energy, while agriculture involves an investment of accumulated energy in a process designed to produce more energy.

Pecuniary economics must take into account the physics of energy as well as the behaviour of men. Money, capital and profit are concepts that relate to man's desires and values insofar as they can be realized through the use of natural resoures. In addition to the values he places on the resources that sustain life and make it comfortable, man assigns value to scarcity, beauty, speed, leisure, power over other men, time etc.

Food, and the fossil fuels are economic sources of energy only if they can be obtained at an energy or work cost that does not exceed the energy or work benefit that can be obtained from them. This fundmental law of energy use defines energy resources. Unless the ratio of benefit to cost, measured in units of energy, is greater than 1:1, the potential resource will fail to become an actual resource. It must be recognized that technological feasibility neither equals nor forecasts economic feasibility. Cook (7) points cut that the prevalent concept that almost anything will become economically possible if enough money is spent on R & D is supported neither by history nor by the logic of physical economics.

The Social Economy of Energy use

The social economy of energy use is concerned with the social costs and benefits accompanying each use of energy, and provides a means of comparing alternative modes and uses. When compared with physical and pecuniary economy, the criteria and measurement of value are more difficult to define for social economy. There are no accepted units of measure, and there are no easy answers to questions of goals and standards. Nevertheless, social economy is of crucial concern, since it is the measure of social adaptation to environmental constraints.

In an industrial society, the direct social costs of energy use will outweight the direct social costs of energy production, because more people are affected by the consequences of energy use than by the consequences of its production.

Ethical and Moral Aspects of Energy use

National and international patterns of energy use raise complex ethical questions, relating not only to their environmental impact, but also to social equity. The moral aspcts of energy use tend to be ignored in economic analysis. Conservationists often refer to our duty to leave a good environment and a stock of resources for future generations.

Pollution was widely regarded as immoral before it became illegal: immoral because it degraded the environment of all for the profit of a few. Currently, depletion of non-renewable resources is becoming a moral issue.

The Role of Renewable Energy Sources

At the outset, it is important to recognize that the pursuit of renewable energy options requires decades of work and dedication. R and D programmes must progress through scientific, engineering; and commercial feasibility before these energy sources can play a significant role. It is important to note that a time window exists for the introduction of these energy Sources. If the new technology is not ready at the proper time to replace competitive matured technologies, the new technology with marginal gains will not achieve acceptance in end use. Careful consideration must be given to the expected features of the competitive technologies at the time the new techology enters the market.

The principal incentive for investigating the renewable energy options is that the fuel cost is very low in comparison to anticipated fossil fuel prices. While these benefits may be offset by high capital costs, these systems are expected to provide protection against rapidly rising fuel prices or an unavailable supply

In the case of solar energy, while the primary energy input is free, the equipment necessary to caputure it and convert it into usable form is not. The highly diffuse nature of solar energy necessitates a large collection area. There are other limitations such as the intermittent nature of source, vulnerability to cloud cover, rain dust pollutions: seasonal variation, etc., which necessitate substantial energy-storage systems. In most cases, in order to ensure reliable and uninterrupted supply, a back-up system incorporating a non-renewable source is also required. Furthermore, the collection equipment, which incorporates a substantial amount of energy-intensive materials, such as glass, aluminium, etc. requires the expenditure of large amounts of highgrade energy derive from non-renewable sources. All these considertions relegate solar energy to a supplementary or complementary role, and expose its "parasitic" nature.

Mohr (9) points out that it is customary for the proponents of alternative energy sources to place their faith in the future: in his opinion however, the age of alternative energy abundance looks remote, at best. He also indicates that solar energy exploitation requires massive amounts of material and even more space. Solar energy systems are constantly exposed to the elements and are susceptible to severe weathering. He points out that the main problem in tapping solar energy lies in converting an erratic, fluctuating energy source into a reliable power supply, which ultimately makes solar power an expensive proposition.

Mohr defines the "yield factor" as the ratio of the amount of energy that goes into the construction and maintenance energy system to the amount of energy that it generates throughout its service life. He estimates the amortization period for solar systems to be 30 years, even under optimal conditions, while the estimated service life is only about 12. This means that while solar energy systems do collect energy, they can never actually pay for themselves. Their fabrication involves a greater consumption of energy, in the form of fossil or fossil fuels, than they can return.

Mohr concludes that the initial expectations for alternative energy sources have been far too optimistic. Given current technological developments and those anticipated in the near future, solar energy can not be regarded as an effective substitute energy for the industrialized nations of the Western World, even though it represents a supplementary form of energy, and needs to be integrated into the existing and reliable conventional energy systems.

Looking to the Future

In as much as oil is relatively scarce commodity of considerable value, the supply of which is dominated by a few producers and for which there are no immediate substitutes in many major uses, Goldemberg et al (5) conclude that oil is simply not a good candidate for meeting the escalating energy demands of developing countries.

While coal reserves are more abundant worldwide than oil, the significant use of coal threatens global climatic change within decades, contributing to the green house effect, through atmospheric build-up of carbon dioxide from fossil fuel combustion.

Apart from enormous costs, widespread use of nuclear power involves risks of radioactives pollution, nuclear weapons proliferation, nuclear backmail by terrorists, etc.

Biomass resources are limited by land-use constraints dictated by the low efficiency of photosynthesis: also, only a small fraction of this resource can be exploited for bioenergy, since forests must serve many purposes, including wildlife habitat.

Notwithstanding the above gloomy scenario. Goldemberg et al (5) point out that it is not necessary to presume that improving human welfare necessarily requires greatly increased energy use. It is necessary to focus on energy uses instead of energy supply. The use of energy is not an end in itself. Energy is useful only in so far as it provides energy services such as cooking, environment conditioning (lighting, heating, air conditioning), refrigreation, mechanical work, and transport.

Alm (10) argues that for the U.S. it is necessary to fully understand the nexus between energy, economic and environmental policy over the next few years. Boosting fuel economy and promoting least-cost electricity seem to serve the interests of economy, energy and environment.

The Role of Technology

Particularly for the developed economics, the doomsday predictions of energy shortfalls leading to a disruption of the present life styles, are expected to be offset through technology innovation. Of all the resources available technology is the only one that new is man-made, and it is hoped that technological frontiers would be opened up for continued social development in both the industrialized and developing nations.

Cook (7) point out that without technology, man was severely limited in the amount of natural subsidy he could draw from nature. He distinguishes between two classes of technology. "Economic Technology" improves the efficiency with which a resource is used, for example, by reducing waste, reducing energy cost of production or by making the product more durable or more effective. It is economic technology which has resulted in improvement of the efficiency of spare heating from 5% of fire places to the 75% of a modern furnace. Increase of the efficiency of fossilfuelled electricity generating plants from 15% to 40% within 50 years, reduction of the cut-off grade for minable copper ore from 5% to 0.5% in 75 years, and a reduction by half the power required for aluminium smelting in 40 years. Economic technology expands a known resource.

Another kind of technology, which he has termed "provident technology", creates a resource from material, energy potential, or flux that did not previously constitute a resoures, or provids a device or system for the conversion or use of the resource that in one stroke creates a vast demand for a resource previously of little use. Examples are the wheel, the sail, the steam engine, the electric, motor, the automobile, nuclear power, etc. It is provident technology that, we hope, will make fusion power, MHD or fuel cells a reality.

Future pressures on technology will include the following demands:

(i) Increase the efficiency of use where possible: in transport, process heating, fuel to fuel conversion, etc.

(ii) Develop renewable energy substitutes for non-renewable energy resources: by making significant advances in breeder, fusion, and solar systems.

(iii) Decrease adverse environmental impact of energy systems at minimal cost in energy and work.

(iv) Develop materials—recycling systems to conserve energy as well as materials.

(v) Push economies of scale to their limits in energy transport (supertankers, ultra high- voltage and cryogenic transmission of electricity, coal slurry pipelines, etc.)

The list is incomplete, and while the consensus goals are obvious, the trade-offs are not. One of the reasons for the non-attainment of the goals is the expansion of man's desires; in addition, technology and innovation have introduced an ever-increasing range of products.

It is to be recognized that human society functions within the purview of a highly complex energy ecology, with both direct and subtle relationships between its internal components, the natural environment and the social environment. Lack of complete understanding the role of energy on the quality of life, ranging from the idealistic notion that the less energy used the better the life, to the opposite extreme or the need for unlimited abundance. Both extremes have little evidence to support their validity.

Starr (3) points out that sufficient evidence exists to show that a significant level of inanimate energy to supplement human labour improves the quality of life. He also proposes that happiness (as it relates to material goods) can be equated to the ratio of consumption to desire, with the ratio always less than one. The traditional approach is to increase consumption through productivity increase. The philosopher's approach is to hope for a reduction in desire.

Goldemberg et al (5) have examined the correlation between per capita energy use and the Physical Quality of Life Index (PQLI), which employes the infant mortality rate, life expectancy, and the literacy rates as measures of well-being. A PQLI of about 90, which is typical for industrialized countries, corresponds to per capita energy use rates of 1.0 to 1.2 KW. Although the index is interesting, there is considerable scatter in the data.

A more fundamental approach is considered to be the estimation of the energy requirements of a future course of economic growth aimed at satisfying basic human needs. In the Latin American world model advanced by the Bariloche Foundation, the optimization criterion relates to the maximization of life expectancy at birth; the assumption being that this criterion approximates the goal of maximising output so that basic human needs are satisfied as quickly as possible. The model also has targets for nutrition, shelter and education.

Some Features of Decision-Making on Energy Matters

Starr (3) points out that all societies seek to maximize net social benefits through the allocation of available resources. The successful deployment of these resources of provide benefits depends on the quality of the decisions made. The key element in these decisions is uncertainty, which is always present. It occurs on three levels:

(i) available options;
(ii) outcomes of the specific options under consideration;
(iii) evaluation of net social benefit.

In the quest for a methodology for interpreting consensus social values relating trade-offs between various social objectives, risk-benefits analysis represents an attempt to provide a method for dealing with some of these trade-offs, those in which risks are significant factors. Three principal areas of application of this analysis may be identified:

(i) Societal choice of expanding or curtailing the development of our existing technological capabilities. E.g. relative emphasis on mass and individual transportation systems.

(ii) Setting of performance targets for existing or new technologies where trade-offs between safety, environmental effects, and cost can produce major social and economic impacts E.g. airquality criteria.

(iii) Decisions between competitive technical systems that produce a similar beneficial function, but with differing social costs. E.g. selection of coal or uranium as the basis for electric power generation.

The hypothesis underlying this method is that the social cost of risk is revealed through a measure of the benefit required for its compensation.

The explanation offered for these behaviors is related to the initiative desire to maximize societal resilience, which is defined as the ability of a society to recover from a damaging event with a minimum of social cost. Some of the factors affecting this resiliency are: the population of the society, population density, resources avilable for disaster response, degree of technological development within the society.

Starr points out that full disclosure of all aspects of decision-making is a key element in assuring pluralistic choices. As an example, the recommendation of the Council on Environmental Quality, that plans for nuclear power development be delayed until ucertainties in radioactive waste management are resolved, is discussed. What was not mentioned by the CEQ was the risk associated with not having nuclear power-whether the risk is due to increased coal combustion, increased imports of energy, or risk from energy shortages. In this context, the following quotation, attributed to Homi Bhabha, is relevant:

"No power is more dangerous than no power"
"No power is more expensive than no power"

It is also pointed out that when individuals "Voluntarily" take risks for personal pleasure or profit, they appear to be willing to accept relatively high risk levels in return for rather modest quantifiable benefits. The situation changes markedly when the individual no longer believes he can control his risk exposure. In such "involuntary" situations, the risk management is in the hand's of a societal group usually remote operationally from the individual and his risk exposure. Under these circumstances, the individual exposed to an "involuntary" risk is fearful of the consequences, and demands a level for such involuntary risk exposure as much as 1000 times less than what would be acceptable on a voluntary basis.

Furthermore, since most new technical system initially appear to be replacements for existing system, e.g. nuclear for coal power, public policy is generally concerned with comparative risk levels rather than comparative benefits.

There is also some confusion in the literature in evaluating the GDP/E ratios: whether it represents the "production" of goods and services and the relative "efficiency" of energy input to the creation of any given GDP, of the relative "consumption" of goods and services and the energy embedded therein.

Starr (3) also point out that energy use in urban areas is likely to be a more effective producer of economic outut. The urban areas have more infrastructure, special labour skills and lower transportation costs for materials. Rural electrification is a luxury that industrialized countries have managed to achieve rather recently. It appears to be economically counterproductive for developing countries. Decentralized or "backyard" energy sources for rural areas can improve comfort and lifestyle, but are unlikely to produce substantial productive increases.

Cook (7) proposes that the social and physical efficiency of energy use coverge in an economy of scarcity, but diverge in an economy of abundance. Thus, where there is widespread unemployment, it may be more efficient socially to use idle hands productively than to use machines efficiently. In an economy of scarcity, the concept of conservation may change from efficient.

Special Characteristics of Strategies for Developing Countries

World energy consumption is dominated by the rich industrialized countries. Even within developing countries, a small percentage of the population representing the privileged classes account for most of the commercial energy use. While petroleum products and electricity predminate among the high-income groups, biomass sources are important for the majority of the population who are poor. While the well-publicized "energy crisis" has been seen almost exclusively as a problem of oil and the other modern energy forms that shape the life- styles of the elites, many among the rural poor face the "other energy crisis" representing the lack of biomass sources for their basic needs of cooking and water heating.

Several alternative development strategies have been proposed for developing countries (5). One alternative was to emphasize employment generation so as to create the purchasing power needed to help the poor. It became apparent soon, however, that poverty in developing countries is rooted not only in widspread unemployment and underemployment, but also in the low productivity of employment.

Another alternative is the basic human needs approach to development, in which societal resources are directed at satisfying a well-defined set of the most basic human needs. The basic human needs approach to development is fundamentally different from simply emphasizing economic growth. It is proposed that by raising the standards of living in rural areas emphasizing basic needs would reduce the pressure to leave the countryside and thereby reduce the need to expand expensive city services.

It is suggested that the strategy for economic development for developing countries should enable them to produce more of what they consume and consume more of what they produce, trading more with other developing countries, and less with industrialized countries. It is contended that a development strategy emphasizing the production of basic materials for export would require far more energy than one emphasizing the production of basic materials mainly for domestic markets. Emphasis on satisfying basic human needs, however, is likely to seek labour-intensive methods of production to generate more jobs for the poor. Many available production technologies are ill-suited to developing countries in that they replace workers with machines wherever possible.

Goldmberg et al(7) raise three pertinent questions relating to the relevance of energy-efficient technologies to developing countries:

(i) Where is energy to save in poor countries, which are struggling to satisfy basic human needs?

(ii) Even if opportunities for savings are significant, since most energy-efficient technology is to be imported by these countries, would it not increase their dependence on industrialized countries?

(iii) Since extra investment is usually required to improve energy efficiency, is energy-efficient technology appropriate for these countries, where capital is scared?

It is strongly suggested that developing countries should not retrace the development path of the North, but should pursue new direction and experiment with promising technologies.

REFERENCES

1 Fellx, fremont, "Growth of Energy Consumption and National Income Throughout the World",IEEE Spectrum, 81-102(July 1964).

2 Darmstadter, J. et al,"Energy in the World Economy: A Statistical Review of Trends in Output, Trade and Consumption since 1925", Johns Hopkins Press, USA, 1971; Cited in Ref.3.

3 Starr,C "Current Issues in Energy", Pergamon Press.1979.

4 Tyner, W.E., "Energy resources and economic development in India" Allied Publishers, 1978.

5 Goldembers,. et al,"Energy for Decelopment". World Resources

Institute. Oxford and IBH Pub. Co. 1987.

6 Stewart, H.B.,"Recollecting the Future", Universal Book Stall. New Delhi, 1989.

7 Cook. E.,"Man, Energy, Society", W.H.Freeman and Co. USA. 1976.

8 Muller, H.J.,"Human Values and Modern Tecnology". cited in Ref 7.

9 Mohr.H."How viable are Alternative Energy Sources?" Suizer Technical Review,4-9 (4,1987).

10 Aim. A.L., "Energy economics and environmental policy". Environ. Sci. Technol., 23 (6), 649 (1989).

CHAPTER 3

Power Generation From Biomass

Dr. B.C.Jain,
Managing Dir., Ankur,
Baroda -390 008

The dismal energy scenarios of the country call for the urgent exploring of all available energy alternatives. The paper lists various attributes that a new energy source should meet to receive favourable consideration for support, at national level. It goes on to show that biomass is indeed one of the most promising energy sources with both short and long term potentials. Biomass resource base is analysed next, in terms of agricultural residues, energy plantation.

Conversion technologies like anaerobic digestion, pyrolysis, combustion and gasification are discussed. Economic considerations for power generation from biomass at various operating levels are next considered followed by a review of likely social and environmental considerations.

1.0 Introduction

The dismal current and future energy scenarios of the country clearly throw up the urgent need for exploring all available alternative energy resources, if industrial and agricultural production as also the entire economic progress of the nation are not to stagnate and suffer. While the energy situation as also the Balance of Payments (BOP) situation is becoming so difficult that almost any new energy source is bound to receive favourable consideration, major impact and contribution by such new sources will require its meeting many of the following attributes :

(i) Given the increasingly large gaps between demand and supply for all forms of energy (i.e. electricity, fuel, thermal energy, etc.), the source should preferably be capable of providing all the energy i.e. electricity, fuel, thermal energy, etc.

(ii) It will also be desirable if the source is available to provide energy/power at both small scale/localised/decentralised level as also at large scale/centralised level (grid connected power generation being an example).

(iii) The source should be capable of providing energy/power "on demand".

(iv) If the source is to start contributing almost immediately, the gestation periods should be short.

(v) If the source is to meet even the peak demand (a major constraint in terms of current supply situation), it should have a very short start-up time.

(vi) Given the tremendous constraints on available financial resources, it should not be very capital intensive.

(vii) As energy/power costs are already seen as being high, cost of producing energy/power should not be exorbitant/very high.

(viii) Given the already precarious environmental and social balances, its widespread use should be environmentally benign/positive and socially desirable.

(ix) Given the increasing concern about global climate (global warming in particular and acid rains etc. in general), its deployment should not result in net CO_2 addition and/or sulphur oxides emission.

It is obvious that any source that meets a large number of these attributes can be excepted to contribute extensively to the national energy scene.

2.0 Relevance of Biomass as an Energy Source

The above listing of the desirable attributes indeed sets up the basic framework for evaluation of the relevance of any new energy source under consideration. It is indeed amazing to note that biomass as an energy resource meets almost all the above attributes, as briefly elaborated below :

(i) Even with readily available conversion technology, biomass can indeed provide all forms of energy (i.e. electricity, fuels, heat, etc.) in an effective and efficient manner.

(ii) By appropriate choice of conversion technologies/devices, it is again possible to meet either small scale, decentralised energy/power needs or large scale, centralised needs (like grid-connected power generation). Of course, for the latter option, biomass availability has to be ensured.

(iii) Because of the natural and cheap storage of energy in the form of biomass itself, requisite energy/power can indeed be made available "on demand". In many ways, biomass as an energy source can be compared with coal/petroleum, except for its relatively low energy density.

(iv) Typical biomass conversion technologies and associated hardware have an implementation time scale of a month to a couple of years (at most) and such short gestation period can indeed allow it to start contributing right away, depending on resources committed.

(v) Meeting peak demands requires a hardware/device which can be started and stopped almost instantaneously. Some of the biomass based technologies (gasification for example) can have very short start up time which make such source-device combination ideal for meeting demand peaks.

(vi) In spite of the various strengths listed above (and others mentioned below), the capital instensity of a number of biomass conversion technologies is quite reasonable and even lower than most of the conventional energy/power production technologies. For example, capital investments for power generation even with currently available technologies are only of the order of Rs.10,000/kw of installed capacity. For the sake of comparison, even a 200 Megawatt thermal power plant requires capital investment of at least 14,000/kw (not counting additional investments in coal mining, railways, etc.)

(vii) Given the low capital intensity, cost of energy/power production is largely dictated by the cost of biomass and, in most instances, unit costs producing energy/power are at least comparable to conventional costs, if not lower.

(viii) Appropriate coupling of energy production programme with biomass production programmes is probably one of the few potential schemes with tremendous environmental and social benefits. Such close coupling could indeed provide the requisite economic reason for afforestation activities to be commercial and successful. Positive environmental impacts of large scale afforestation are too well known to require mention here and very large scale employment generation for relatively weaker sections of the society can also be ensured through such production programmes.

(ix) Once biomass production and its conversion for energy purposes are closely coupled, there is absolutely no net CO_2 release and given the very nature of afforestation and plantations, the net impact on CO_2 can be easily expected to be in terms of a slight decrease. This aspect is also very well recognised the world over and this has indeed been recommended as one of the few options available to mankind for arresting global warming. Of course, biomass has almost negligible sulphur content and hence problems of acid rain etc. can also be helped.

An indepth analysis therefore clearly shows that biomass is indeed one of the most promising energy sources with both immediate gains as also extensive long term potential. The low level of technological development today is a result of historical reasons wherein undue emphasis was put on the more convenient and exotic sources like petroleum products, nuclear energy) without adequate consideration having been given to associated problems of these sources. As the realisation of these associated problems (which are both immediate and long term in nature) sets in, greater reliance will have to be placed on various new and renewable sources in general and biomass in particular. It would seem apparent that substantial technological input could indeed result in great convenience and reliability being built into various biomass conversion devices.

3.0 Likely Biomass Resource Base

While the term "biomass" covers a wide spectrum of materials, three major categories are considered here as these are expected to be indicative of the overall potential. The categories are :

(i) Agriculrural Residues
(ii) Agro-Industrial Residues
(iii) Energy Plantations

Other likely contributors to the overall resource base are weeds and annual sustainable yields from existing forests (densely forested areas like Bastar in M.P., A & N Islands, some of the north eastern states, etc). However, the relative contribution to energy and power through these weeds and forested areas is likely to be small in the overall scenario for a variety of reasons though the possibility of significant local impact should not be ignored.

3.1 Agricultural Residues

The current resource base used as a reference is from the recently conducted (1984-86) National Productivity Council study titled "Improvement of Agricultural Residues and Agro-Industrial Byproducts Utilisation". Sponsored by the Department of Non-Conventional Energy Sources, the study covered fifteen major states i.e. Andhra Pradesh, Bihar, Gujarat, Haryana, Himachal Pradesh, Karnataka, Kerala, Madhya Pradesh, Maharashtra, Orissa, Punjab, Rajasthan, Tamil Nadu, Uttar Pradesh and West Bengal. The major conclusions of the study regarding agricultural residues were :

(1) Estimated total production of agricultural residues in the country was close to 320 million tonnes per annum (a level which is much higher than even the current consumption of firewood and agricultural residues put together).

(2) The major crop residues were paddy straw, wheet straw, jowar straw, sugarcane trash, groundnut holms, maize stalks and bajra straw. These residues accounted for almost 88% of the total residues in the country.

(3) Prominent among the states contributing most to the production of residues are U.P., Punjab, M.P., A.P., Bihar, West Bengal, Maharashtra, and Tamil Nadu which together accounted for almost three quarters (73.6%) of the total residues.

(4) Seasonally, almost 60% of the residues were seen to be produced during October to December and another 25% during April to June.

In terms of resource base, while almost all biomass can be used for energy purposes, such a gross analysis is only of academic value and many of the residues have economically and socially more attractive uses other than as fuel. However, even if a fraction is available for energy purposes, the resource base can be seen to be quite large.

3.2 Agro-Industrial Residues

The base line data have once again been taken from the NPC study (mentioned in the earlier section) which covered agricultural residues, agro-industrial residues and animal dung as resources. The major agro-industrial residues covered for the fifteen major states listed earlier included bagasses, rice husk, molasses, groundnut shells, saw dust and cotton gin waste. A few of the salient observations of the study were :

(1) The annual production of by-products or agro-industrial residues covered by the study was over 50 million tons in the country in 1985-86.

(2) Major contributing states were U.P., A.P., Tamil Nadu, Maharashtra, Karnataka, Punjab, and Gujarat which together, accounted for 78.6% of the total by-products available in the country.

(3) Contribution of bagasse and rice husk put together is more than 86% of the total.

It may be noticed here that most of the by-products are even today used as fuel. However, in many instances, the efficiency of utilisation is indeed quite low and the typical applications are for low grade thermal heat - both the facts which could be easily changed using gasification. In this context, the resource base is again very significant.

3.3 Energy Plantations

A concerted programme of demonstration energy plantations has been launched by DNES during the seventh five year plan in light of the wide spectrum of benefits likely to accrue from such plantations. The plantations are being undertaken on problem soils, i.e. saline/alkali soils, water logged soils, ravine lands, shifting sand dunes etc. Given that the purpose is to produce energy rather than industrial wood, the species are selected to achieve the following :

(1) High biomass yield,

(2) Biomass of high calorific value,

(3) Fast growth rate,

(4) Good coppicing ability (i.e. regeneration after harvesting),

(5) High adaptability and survivability in difficult soils with low rainfall,

(6) Ease of establishment of the plantation,

(7) Very little after-care.

While a number of plantations have already been taken up by the department during the last few years, the basic resource for the purpose of this report has to be in terms of wastelands that could be made available for the purpose of raising such plantations. Table 1 gives the potential land area that could be available for this purpose in different states and regions and it works out to 62.5 million hectares.

Given that typical current energy plantation yields range from 5 tons per annum per hectare to 20 tons per annum per heactare, the planned resource production could indeed be very large.

4.0 Contribution Potential

Contribution potential of the biomass resource (and, hence, gasification technology) in the larger national context could be seen from a number of angles. One possible comparsion is with the current consumption levels of non-commerical energy sources like firewood and agricultural residues wherein it can be clearly seen that the potential is much larger than the current and projected consumption levels of above. Another more relevant consideration for a futuristic technology like gasification is in terms of its power generation potential. This has been attempted recently in a DNES study entitled "Techo-Economic Studies in Technological Options for Power Generation from Biomass", with very conservative assumptions, the study indeed arrives at rather revealing conclusions which are briefly mentioned here.

Overall assessment of ultimate, theroretical power generation potential from biomass is summarised in Fig. 1. It must be clearly mentioned here that the potential shown here basically is of academic interest only as it assumes that all available biomass can, in the long run, be allotted for power generation. While wasteland has been used as the base for energy plantation component, residue production for year 2000 has been used as the base for both agricultural residues as

Table 1

Land Considered As Being Available For Fuel Wood/Fodder Plantations (million ha)

State	*Open/ Degraded Forest Land (1)*	*Cultivable Waste Land (2)*	*Barren Land Uncultivable Land (3)*	*Permanent Pastures & other Grazing Lands*	*Farm Perimeter Land*	*Total*
Northern Region						
1 Uttar Pradesh	0.290	1.338	1,146	0.298	0.937	4.009
2 Rajahthan	0.270	0.880	2.960	1.834	0.469	12.413
3 Punjab	0.026	0.048	0.099	0.003	0.111	0.287
4 Haryana	0.026	0.036	0.073	0.043	0.084	0.262
5 J & K	0.140	0.149	0.330	0.124	0.049	0.692
6 Himachal Pradesh	0.240	0.136	0.356	1.024	0.039	1.595
Eastern Region						
7 West Bengal	0.120	NA	1.352	NA	0.257	1.759
8 Bihar	0.640	0.468	1.008	0.144	0.571	2.831
9 Assam	0.420	0.130	1.533	0.185	0.090	2.358
10 Orissa	1.040	0.260	0.279	0.534	0.266	2.379

(Contd.)

Southern Region							
11	Andhra Pradesh	0.840	0.888	2.134	0.948	0.419	5.229
12	Karnataka	0.470	0.530	1.051	1.398	0.293	3.742
13	Tamilnadu	0.370	0.362	1.678	0.161	0.373	2.944
14	Kerala	0.110	0.123	0.260	0.006	0.121	0.620
Western Region							
15	Madhya Pradesh	2.130	1.849	2.303	2.877	6.613	9.772
16	Gujarat	0.220	2.000	2,486	0.850	0.233	5.789
17	Maharashtra	1.116	1.020	1.725	1.592	0.491	5.944
	Total	8.468	16.217	20.503	12.021	5.416	62.625

(1) For the years 1980-82, Source: National Remote Sensing Agency, Department of Space, December, 1983.
(2) For the year 1978-79: Source: Indian Agriculture in Brief: Ministry of Agriculture.
(3) For the year 1978/79: Source: Indian Agriculture in Brief: Ministry of Agriculture.

Overall Source : "Towards a Perspective on Energy Demand & Supply in India in 2004/05." Advisory Board on Energy, Govt. of India, 1985.

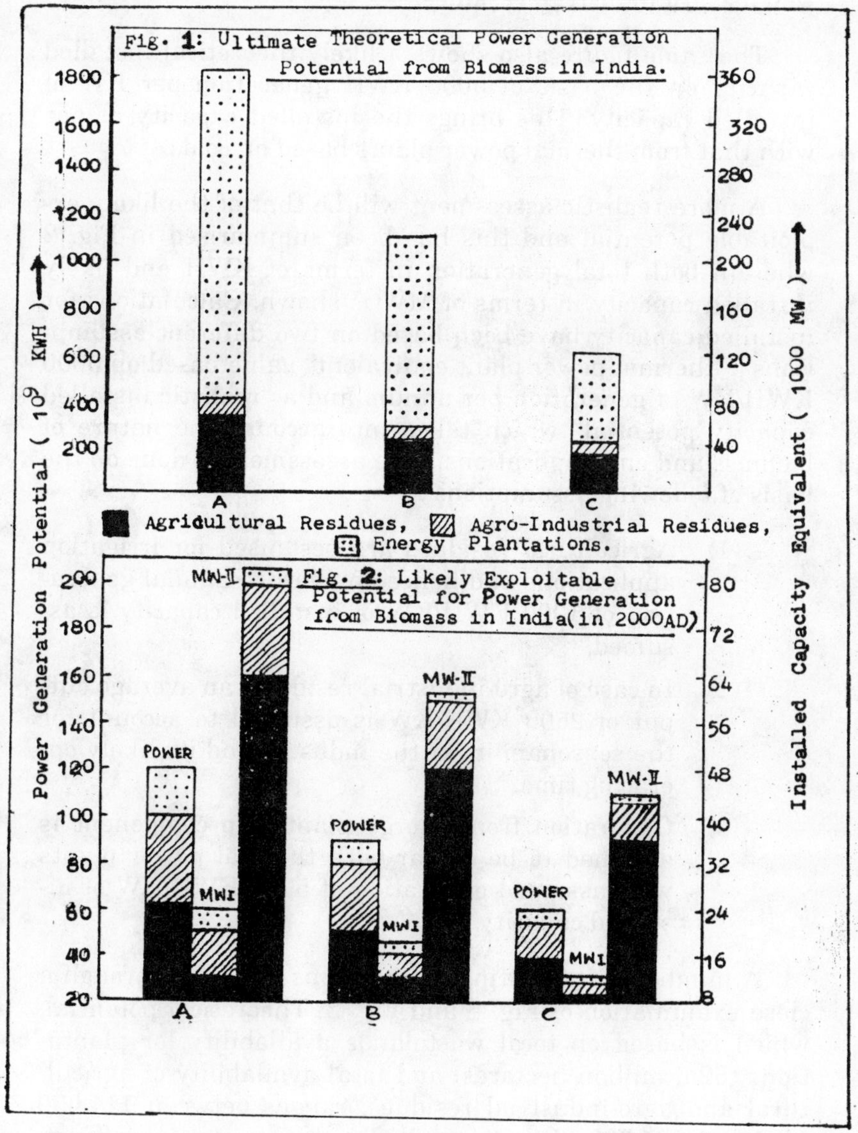

A: Optimistic Projection, B: Likely Projection, C: Pessimistic Projection

MW-I : Installed Capacity—Thermal Power Plant Equivalent

MW-II : Realistic Installed Capacity.

well as agro-industrial residues.

The same figure also shows a likely theoretical installed capacity on the basis of 5000 KWH generation per KW of installed capacity. This brings the installed capacity at par with that from thermal power plants based on coal.

A more realistic assessment will be that of the likely exploitable potential and this has been summarised in Fig. 2 wherein both total generation in terms of KWH and likely installed capacity in terms of MW is shown. Calculations for installed capacity have been based on two different assumptions - "thermal power plant equivalent" value based on 5000 KWH/KW of generation per annum and a "realistic installed capacity potential" which takes into account the nature of biomass and end applications. This assessment is done on the basis of following assumptions :

(1) Agricultural residues are best used for irrigation applications and hence an average annual generation of 1000 KWH/KW of installed capacity is assumed.

(2) In case of agro-industrial residues, an average output of 2500 KWH/KW is assumed to account for the seasonability of the industry and its likely operating time.

(3) Generation from energy plantation component is assumed to be at par with thermal power plants with assumed generation of 5000 KWH/KW of installed capacity.

A number of interesting observations come out through a close examination of Fig. 1 and Fig. 2. Theoretical potential, which is based on total wastelands availability for plantations (62.6 million hectares) and total availability of agricultural and agro-industrial residues, ranges between 114,000 MW and 365,000 MW of installed capacity-staggering numbers indeed. The basic purpose of such numbers is to clearly bring out the relevance of the resource - the fact that even if a small fraction of the ultimate theoretical potential is exploitable and only a part of the latter is actually put to use, the contribution to the national energy scene could still be quite significant.

Fig. 2 brings out precisely such significant potential for contribution. Even in the most pessimistic case, likely exploitable power generation pottential for biomass by year 2000 could be as much as 61 x 10 kilowatt-hours of annual generation - equivalent to more than 10,000 MW of installed capacity in terms of thermal power plants. Of course, considering that many of the biomass-based power generation systems will be for irrigation, decentralized power generation, captive power generation, etc. with a much lower plant load factor, the actual exploitable installed capacity even in the most pessimistic case could be more than 44,000 MW. It should be quite clear that even if a part of its is indeed employed by year 2000, it will have an impact on the national scene.

The more likely exploitable potential for power generation through biomass by the year 2000 is 86 x 10 kilowatt-hours of electricity per annum, more than 17,000 MW of equivalent installed capacity and as much as 61,600 MW of actual installed capacity. Even without looking at the more optimistic projection, the real possibility of discernible impact clearly comes through. In all cases, the main contribution is from agricultural residues followed by agro-industrial residues.

In a nutshell, it should be clear that power generation potential from biomass is not likely to be limited on the basis of the resource base. It is important to recollect that the only resources considered in "likely" and "pessimistic" assessments have been 1.0 million hectares of wastelands (out of total of 62.6 million hectares), agricultural residues which are already partly being used for energy purpose (cotton sticks, tuver stalks, part of sugarcane trash, etc. and even a smaller share of paddy straw, jowar straw, etc.) and agro-industrial residues like bagasse, rice husk, groundnut shells and wood waste and saw dust.

5.0 Biomass Conversion Technologies

Though biomass conversion technologies and devices cover a very wide spectrum, these can be covered generally through four major categories.

(i) Anaerobic Digestion,
(ii) Pyrolysis,
(iii) Combustion, and
(iv) Gasification.

Anaerobic digestion has been extensively used in the country for high-moisture organic materials, particularly dung. Major application has been for domestic and institutional cooking though a few installations for industrial heat as well as shaft power/electricity have also been made. The basic process is also used for treatment of industrial and municipal wastes and effluents with a number of installations for distillery effluent and human excreta having been made in various parts of the country. Being based on bacterial activity, the process has a long cycle time (running into 4 to 6 weeks at times) and the long retention times and high moisture/water content also result in relatively large plant sizes with associated capital intensity. Total energy availability on a daily basis is obviously restricted to the average gas production levels and additional "on demand" generation cannot be attempted (even if additional feed is available) because of long retention/cycle times.

Pyrolysis could be largely used for production of various gaseous, liquid and solid fuels from biomass, with the most common example being conventional charcoal making. Though the conventional technique suffers from low energy efficiency and high environmental pollution, well designed charcoal kilns can overcome these drawbacks. Production of liquid fuels etc, through pyrolysis, has not been attempted on any significant scale.

Biomass combustion can be said to be one of the oldest technologies known by mankind - beginning with the first fires built by man. Industrial revolution was largely based on combustion and biomass combustion has been widely practised all over the world for producing heat as also for generation of shaft power/electricity.

6.0 Biomass Gasification & Its Special Relevance

Biomass Gasification is basically conversion of solid biomass (i.e. wood/wood waste, agricultural residues etc.) into a combustible gas mixture normally called "Producer Gas" (or Law Btu Gas). The process is typically used for "woody" biomass and it involves partial combustion of such biomass. Partial combustion process occurs when air supply (or more precisely, oxygen) is less than adequate for combustion of biomass to be completed. Given that biomass contains carbon, hydrogen and oxygen molecules, complete combustion would produce carbon dioxide (CO_2) and water vapour (H_2O). Partial combustion produces carbon monoxide (CO) as well as hydrogen (H_2) which are both combustiblee gases.

Solid biomass fuels are usually inconvenient, have low efficiency of utilisation and can only be used for certain limited applications. Combustion is the normal conversion process and while direct thermal use in cooking, heating space and water or generation of steam is possible (usually with low efficiencies), generation of power, for example, requires high/medium pressure steam boiler alongwith steam engine or turbine with accessories. For small power needs (a few kilowatts to a few hundred killowatts), this conversion technology is not only capital intensive and complex, it is also very inefficient.

Conversion of the same biomass to a combustible gas mixture like producer gas removes most of these problems with use of solid biomass fuels. While conversion to gas results in loss of energy of upto 25%. use of gas can be highly efficient and hence overall efficiency could be very high. Furthermore, the gas can be directly fed into internal combustion engines (I.C. engines) wherein it could save precious petroleum fuels. Also, it can be employed at any scale (above a few kilowatts) and hence is ideally suited for decentralised applications - whether for shaft power, electricity or thermal energy.

In case of shaft power/electricity, the gas is basically burnt inside an engine. The engine can be either diesel based (i.e. compression ignition or C.I. engine). Given that the producer gas-air mixture does not auto-ignite at normal engine pressures and temperatures, running diesel engine solely on gas is not possible and pilot diesel injection is needed to start combustion. In case of S.I. engines, spark plug initiates combustion and hence such engines can be run totally on gas. Wherever diesel engines are used, "diesel replacement" therefore becomes a key parameter as economics largely depends on what percent of total normal consumption can be replaced by gas.

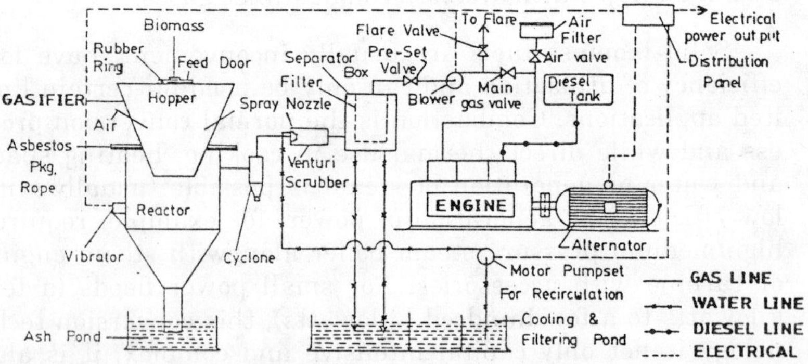

Fig. 3 Ankur Biomass Gasifier-Schematic

A very important advantage of biomass gasification is the fact that it can be employed equally effectively at both small scale, decentralised level as well as large scale, centralised level.

Special relevance of biomass gasification can be further highlighted by briefly listing the range of applications, certain useful numbers as also by looking at just one example i.e. irrigation water pumping.

1. Range of Applications

- Thermal Applications
- Simultaneous Charcoal and Power Production
- Irrigation Pumping
- Pumpset Energisation/Village Electrification
- Captive Power for specific industries
- Grid-fed power from Energy Plantations on wastelands.

2. Some useful numbers

- 3 to 4 kg of biomass save a litre of diesel
- 65% to 85% replacement of diesel in C.I. Engines
- 100% gas operation with S.I. Engines
- 5 hp pumpset operating 1000 hours needs only 3.5 tons of biomass
- 1 MW for 4000 hours requires 4000 tons of biomass
- For irrigation, 3-4% of land for tree growing/25% of residues
- Typical plantation yields are 10 to 60 tons per ha.

3. Application Details - Irrigation Pumping as an example

- 10 million pumpsets
- 40% diesel based
- Annual production of diesel engines-half a million
- Narrow power range of 3.5 kW to 7.5 kW
- Reasonably steady load
- Insensitivity to minor speed fluctuations
- 3-4% of land for trees/25% of residues can support perennial irrigation

Even with relatively low technological development as of now, it is quite evident that biomass gasification as a technology has very special relevance to a country like India.

7.0 Power Generation from Biomass - Economic Considerations

Commercially available biomass gasification systems cover the range of 3 kw to 100 kw (as of 1990), with likely extension of the range to single units of 500 kw before the end of 1991. With such hardware availability, immense possibilities are opening up for power generation through gasification, using various residues and wastes either in raw form or after charring/briquetting etc. The major relevant power generation categories are :

Decentralised / Captive Power Generation

This category also covers a large number of possibilities and some of these are :

(1) Energisation of a number of pump sets located in a cluster

(2) Captive power for industries that have biomass waste process (Saw Mills, Joinery Mills, Paper Mills etc)

(3) Captive power for industrial units located in rural areas and having extensive land areas which can be devoted to energy plantations, and

(4) Village electrification in remote rural terrain which have adequate biomass resources.

Large Scale Power Generation for Grid Feeding

Energy plantations are increasingly being taken up by various State Agencies as also by interested farmers and business houses. Such plantations are typically based on fast growing tree species, many of which are of the coppicing variety. Biomass produced from such plantations can be effectively converted to power on a perennial basis and this power can then be fed into grid for general use.

Table 2

Economics of Power Generation From Biomass - A Summary

Sr No.	*.Application*	*Conversion Technology*	*System Rating (KW)*	*Assumed Biomass Price (Rs/Ton)*	*Cost of Genera tion (Rs/k Wh)*	*Conventional Option & Typical Cost (Rs/k Wh)*	*Remarks*
1.	Village Electrification	Biomass Gasification	24 KW	Rs.300 & Rs.600 per ton	1.57 to 1.90	*DG Set:* 1.98-2. 02 *Grid Extension* 1.90-2.10	Two different Interest rates (9% & 12%) have been used
2.	Pumpset Energisation	- do -	16 KW	- do-	3.30 to 3.82	*Diesel Pumpsets* 2.02-2.12 *Grid Extension* 2.71-3.01	1. Same as above 2. Internal grid for 8 pumpsets included.
3.	Capative, Industrial Power Generation	- do -	300 KW	- do -	1.20 to 1.54	*DG Set* 1.61-1.65	No.1 as above

(Contd.)

4.	Power from Energy Plantatiens	- do -	2000 KW	Rs.220 & Rs.400 per ton	0.51 to 0.92	*Thermal/Hydro/ Nuclear Power* 0.60-1.00		- do -
5.	- do-	Dendr-othermal Power Generation		1000 KW	- do - 1.56	1.10 to	- do -	- do-

Extensive economic analyses for power generation from biomass have been carried out, with a number of case studies covering various applications listed above. Equipment costs have been based on actual market prices while sensitivity analyses have been carried out for other relevant parameters.

A summary of the results obtained for various cases is in order for an effective review of the current economic viability of technologies for power generation from biomass. Such a summary is given in Table 2. It clearly brings out the economic viability of biomass based power generation even at today's technology level. Almost all competing conventional options (mainly electricity from thermal or hydro plants and diesel engine pumpsets/gensets) have reached a relatively mature stage of technological development while the field of biomass combustion and gasification etc. has, so far, received scant attention. Even if nuclear energy is considered alongwith biomass-based technologies, the developmental effort that has gone into nuclear power related activities is many orders of magnitude higher than the effort in biomass based technologies. It is, therefore, reasonable to assume that a concentrated and sustained developmental effort could significantly reduce investments, improve operating efficiencies and result in systems that may have excellent user acceptance. All of this can only improve the economic competitiveness of these technologies further.

Another major consideration in case of biomass based power generation is the cost at which biomass can be made available for the purpose. In many examples, the cost has been taken to be as high as Rs. 600 ton which reflects the prevailing market prices in many of the biomass - scarce regions. It is to be pointed out that any major expansion of biomass based power generation should automatically include specific biomass production programmes to provide the requisite feedstock. As has been conclusively shown in case of energy plantations on wastelands etc., the cost of planned biomass production is much lower than the scarcity price of Rs. 600 per ton. Once again, such a production programme can significantly reduce costs of generation making such power even cheaper compared to other established, conventional means.

8.0 Social Considerations

Three major components of any significant biomass based power generation programme will consist of agricultural residue utilisation, agro-industrial residue utilisation and wastelands utilisation for power generation based on energy plantations. A well-conceived programme is likely to have multiple positive social impacts with a variety of synergistic effects. Such a programme will require judicious coupling of biomass production and conversion programmes to ensure that the currently available scarce resources (particularly woody biomass in many parts of the country) are not stretched further to support new conversion technologies resulting in deprivation of the poorer sections in rural areas of the country. This one single action is crucial in eliminating all negative social impacts and the specific benefits described below indeed assume that significant extension and expansion of power generation technologies based on biomass will only follow implementation of an appropriate biomass production programme. Of course, a small technology demonstration programme adding up to an installed capacity of a few hundred megawatts and spread all over the country is unlikely to have any perceptible advers impact and, hence, can be undertaken as a preparatory phase for a much larger programme. Also, bigger programmes can be undertaken in biomass - surplus areas (examples being Bastar in M.P., A & N group of islands, north-eastern states, etc) with appropriate built-in safeguards.

Of the many positive social benefits, a few are considered in some detail here. These include:

(i) Employment generation in economically backward areas;

(ii) Likely favourable impact on social inequities;

(iii) Savings of foreign exchange.

Both production and utilisation of biomass are largely labour intensive technologies. A specific example of energy plantations is taken here to bring out the vast employment potential that these technologies generally have. Experience with DNES-funded energy plantation activities in Gujarat has shown raising of energy plantations on wastelands typically involves anywhere between 300 and 600 man-days of labour per cycle for every hectare of land depending on the intensity of irrigation, nutrient inputs etc. Additional labour is required for harvesting of the trees, their transportation as well as final sizing etc. for its use in various conversion devices. Operation and maintenance of the power generation system is the next phase requiring both skilled and unskilled labour inputs. Of course, developmental work on such conversion technologies and devices is likely to create opportunities for the technically qualified and scientifically oriented professionals. The manufacturing sector is another employment avenue with significant potential because of the very nature of technologies and devices.

Even if only employment opportunities for the unskilled, rural labour (mainly landless labour, marginal farmers etc.) are considered, the total employment generation per hectare of wasteland put to energy plantations works out to be between 400 and 800 man-days per cropping cycle which could typically be of the order of 5 years or so.

Most of the action in terms of biomass production, transport and processing/preparation is likely to take place in rural areas which will be a step in the right direction in terms of prevailing rural-urban economic disparities. And lastly, given that the major chunk of work creation is going to be in the unskilled category, the favourable impact can again be expected to be oriented to the weakest of the weak in the social order. Of course, some imagination and innovative time scheduling will be necessary to properly spread the employment over the year to ensure that no undesirable conflict with agricultural operations is created. Depending on the extent of employment generation in a given areas because of the biomass programme, there could even be a further beneficial impact in terms of ensuring automatic payment of minimum wages because of the modified labour supply-demand situation which may be more favourable to the unemployed/under employed labour force.

The third important benefit to the society and to the nation as a whole is in terms of foreign exchange savings that could be effected by large-scale employment of biomass based power generation technologies for saving imported petroleum products like HSD (high speed diesel) etc. Major contributors to such fuel savings are likely to be systems used for irrigation water pumping, village electrification and captive power generation - applications where diesel pumpsets or gensets could be the conventional options. In terms of potential, the likely impact by the year 2000 could be equivalent to at least 10,000 MW of thermal power plant equivalent capacity (i.e. two-thirds of the contribution expected in year 2000 from agricultural residues and agro-industrial residues). Though 100% gas engines are likely to be in vogue by that time, even at 80% diesel replacement, the likely annual diesel savings for the year 2000 work out to about 10 million tonnes and the foreign exchnge savings even at today's prices could be close to Rs.2,400 crores.

There are a number of other indirect social benefits in terms of improved environment, climate, etc. and the impact such improvements could have on health and general well-being of the local population. Some of these benefits are covered in the next section

9.0 Environmental Considerations

Before focussing on the various positive environmental impacts that a properly implemented biomass programme is likely to have, a few potential hazards need to be clearly mentioned. As suggested earlier, any large scale extension and expansion (beyond a small, few-hundred megawatt preliminary demonstration programme) of biomass based power generation programme must only follow an effective biomass production programme to prevent further stress on the limited forest cover and greenery that the country has. This aspect is very well understood by almost everyone concerned with any aspect of biomass programme and, given the current environmental awareness, it can be expected to be properly taken care of.

The only other aspect that needs careful analysis is the possibility of some of the conversion technologies having negative envirnomental impacts. The only serious concern in this regard would seem to be the tars typically produced during biomass gasification which may be carried by the process water if wet scrubbing is employed in the system. Fortunately, removal of tar and neutralisation of process water are not only technically easy, these do not impose any serious economic penalty on the system. As a matter of fact, tar recovery could be, both economically and on energy considerations, quite attractive particularly for larger systems. The need therefore is for awareness and adequate caution so that the only possible negative impact is totally avoided from the very beginning.

The positive environmental impacts of a significant biomass based programme can once again be assessed through a close look at energy plantations on wastelands. A typical biomass production and utilisation cycle will always leave at least four-fifths of the total planted area under green cover. As a matter of fact, the basic national policy of increasing forest cover from the current 22% to a desirable 33% is indicative of the desirability and necessity of quickly increasing such green cover. The impact is even more significant for states and geographic areas that have a much lower green cover (Gujarat, for example, has only 6% effective cover) and, usually, a much higher waste land area. As most of the positive impacts are quite well known, no detailed analysis is attempted and the major aspects are briefly listed below. The listing also includes other potential environmental and climatological benefits from biomass based programme in general.

- The biomass programme can be expected to have a significant positive impact on the rainfall pattern which has been causing serious concern during the last few years.
- It can also be a major element in a strategy for arresting soil erosion - a problem being faced in almost all parts of the country, but more so in the hilly regions.

Another associated benefit is in terms of reduced danger of land slides, etc.

- Rainwater retention and recharging of underground water resources assume an important character in a country like India where the entire fresh water is received over a short time period during monsoon. Biomass programme can be of assistance in this area as well.
- A significant biomass programme is likely to result in a living environment of a much higher quality due to reduced air and water pollution as also due to improved health and general well-being of the local populace.
- The very low ash content of most of the biomass species is another signicficant gain, particularly because of the very high ash content of Indian coal. Increased particulate loading of air can be reduced and the ash disposal problem can be partly assuaged by a large biomass programme. The scars on land due to coal mining need to be seen in the coal belt to be believed. So is the environmental pollution - both of air and water. A significant biomass programme can partly reduce coal requirements and, hence, reduce the land, air and water pollution in mining areas.
- Increasing levels of carbon dioxide in the atmosphere due to greatly increased fossil fuels use are now seen to be resulting in the general warming up of the earth's atmosphere with colossal long-term implications. Such adverse climatological changes can be addressed partly through increased green cover. Also, biomass utilisation essentially recycles CO_2 without introducing additional amounts as fossil fuels do.

From all of the above, it is reasonably clear that the social and environmental considerations are largely favourable for a large biomass production and utilisation programme and, while quantification in financial terms is likely to be complex and difficult in many instances, the overall cost-benefit analysis is bound to be much more favourable for the biomass programme than what is shown by a strict financial analysis.

CHAPTER 4

Power Production Imbalances in India: Its Genesis and Need for Rectification*

A.N.Singh
Former Chairman, CEA and ex-officio Secretary to Govt. of India.

India has had a tremendous development of electric power generation and supply facilities-something like 45 times-in the past year-of which the country can be genuinely proud. This development, however, is unfortunately heavily imbalanced so that large segments of people in many area have comparatively very little electrical energy at their disposal. This lacuna has arisen mainly because, there has been no machanism in the planning process to take notice of growth of such an imbalance, even though it is well known that imbalance in electricity is both the cause as well as indicator of overall economic/social imbalance.

The article draws attention to the magnitude of imbalance that exists in various parts of this country, to show that this discrepancy has been growing, that it is not self correcting and cannot get corrected by itself even in any length of time. Consequently, it is necessary to take specific steps to correct this position.

---

* Reproduced, with kind permission, from Urja, Vol. 30(6). December 1991.

Some suggestions for corrective action are offered on the analogy of methods that had earlier been used in other parts of the country with very good results. An earnest plea is to be made for immediate and continued attention to balancing the growth of electricity supply in this country for a harmonius future.

Introduction

It is readily accepted by all that electricity is a vital input for development. This was well recognized in this country even before independence and it was no surprise that highest priority was devoted to its development since independence. Electricity development has received the highest single segment of Plan funds, of the order of 20% from the very beginning. This attention has borne fruit and the electricity supply facilities have grown by over 45 times in 45 years - showing the highest rate of growth of any segment of the economy. The electricity generating capacity stood at over 65,000 MW in March 1990. A large network of 400 KV transmission lines has come up and even HVDC transmission has been recently introduced. This creditable development of electric power has formed a basic input for the excellent development in the country in industry, agriculture, food grains, transportation etc.

The reasons that apply to high priority for electricity development for the country as a whole, also apply with equal force to equitable and well balanced development of electricity in all parts of the country thereby enabling a similar pace of development of economic activities everywhere. Conversely, an imbalance in the matter of electricity could be indicative of wider imbalance and should induce thinking of corrective steps needed for all concerned areas of activity. Unfortunately however, there has been no practice of reviewing the balance or imbalance of development of electricity in different parts of the country. There by a very important feature in the country's holistic development has gone unattended.

As a result, major imbalances of development of electric power have come up which need urgent corrective action which also as yet needs to be thought out. Even while considering whole states, the ratio of per capita power use stands as large as 1 is to 8 between the least developed and most developed.

This chapter with the objective of highlighting urgency of corrective action makes a review of the development of electricity and electricity production facilities in various parts of the country and brings out a factual picture showing the existing differentials of development. Factual data gathered relating to electric power available for use by the population in various areas of the country is first presented. The data is then discussed from the view point of the magnitude of the relative power availability in the different areas. The aspects of the causes for low development in some places has to be considered - whether it is due to low generating capacity or due to inefficient use. Finally, suggestions for rectifying the imbalance are given. As will be brought out in the detailed discussions that follow the magnitude of differences in electricity availability are vast and the need of corrective action is urgent.

Usual reviews of electricity supply that are taken either consider each state by itself or alternatively are made for the country as a whole. No attention is given to seeing as to whether one state is lagging too far behind while others are running ahead and thereby induce thought for doing something corrective. A rather amusing result of this country wide method of review is that gross shortages of a very badly off part of the country gets averaged out as a shortage for the better off parts.

The discrepancies in level of power supply in different areas have today grown to very large magnitudes. And a thing which should be a matter of great concern is that the existing planning and implementation processes do not have any inherent self correction feature which would rectify the situation. These existing procedures would in fact continue to widen the gap. It will here be relevant to recall that in the last ten years the country has given attention to similar discrepancies and aberrations which have emerged in the process of growth and evolved various positive measures such as industries in no industry districts, etc.

The differentials in the matter of a basic infrastructure like electricity, in a similar way, needs to be assessed and corrected with highest priority.

Present picture of electric power use in different parts of India

The vast development of electric power that has taken place in India is highly uneven-some places (major states) have as much as 5 to 8 times more of it than is available to poor off states. The position becomes evident when normated data relating to power use is seen. Statistics normally published relating to power use gives information such as total installed capacity or total power generation, etc. of different states. However since the states are of widely different magnitudes and not comparable in size of population, such total statistics is not useful for study of relative development. It is essential to normalise the information in an appropriate way.

A very simple way of such normalisation is to use the statistics of per capita consumption of electricity of states. This data directly corelates the production and supply facilities of the state and its population. The data of per capita consumption is available from official publications and forms an authentic basis. It is to be remembered that since electricity can neither be stored nor exported, the consumption perse is representative of the capability of production with due allowance for losses.

The data of per capita consumption of electricity for 20 principal states of the country has accordingly been gathered for 3 sample years covering a span of past 10 years-viz 1979-80, 1985-86, and 1989-90. Care has been taken to see that the date is reliable.

This data on per capita consumption is set out in columns 2, 3 & 4 in Table 1 below.

In columns 5,6, & 7, the same data is presented in a relative form, i,e., the consumption in each state expressed as a percentage of the highest per capita consumption in any

Table 1
Per Capita Electricity Consumption in States

State	*Per Capita Elec.Energy Consumption (KWH/YR)*			*Per Capita Energy of Cols.2,3,4"As % of the Maximum Value.*			*Percent Growth*	
	1979-80	*1985-86*	*1989-90*	*1979-80*	*1985-86*	*1989-90*	*1989-90 / 1985-86*	*1989-90 / 1979-80*
(1)	*(2)*	*(3)*	*(4)*	*(5)*	*(6)*	*(7)*	*(8)*	*(9)*
Punjab	314	423	639	100.0	100.0	100.0	57.0	103.5
Maharashtra	226	313	405	71.9	74.0	63.4	29.4	79.2
Gujarat	243	299	399	77.4	70.7	62.4	33.4	64.2
Haryana	202	247	353	64.3	58.4	55.2	42.9	74.7
Tamil Nadu	181	213	295	57.6	50.4	46.2	38.5	63.0
Karnataka	146	187	273	46.5	44.2	42.7	46.0	87.0
Andhra Pradesh	98	183	227	31.2	43.3	35.5	24.0	131.6
Madhya Pradesh	96	168	205	30.6	39.7	32.1	22.0	113.5
Rajasthan	101	140	202	32.2	33.1	31.6	44.3	100.0
Orissa	110	130	181	35.0	30.7	28.3	39.2	64.5

(Contd.)

(1)	(2)	(3)	(4)	(5)	(6)	(7)	(8)	(9)
Him, Pradesh	57	123	180	18.2	29.1	28.2	46.3	215.8
Jammu & Kashmir	73	116	178	23.2	27.4	27.8	53.4	143.8
Kerala	99	140	164	31.5	33.1	25.7	17.1	65.6
Uttar Pradesh	87	118	159	27.7	27.9	24.9	34.7	82.7
West Bengal	114	135	139	36.3	31.9	21.8	3.0	21.9
Meghalaya	32	76	108	10.2	18.0	16.9	42.1	137.5
Bihar	81	95	102	25.8	22.5	16.0	7.4	25.9
Manipur	6	32	80	2.0	7.6	12.5	150.0	1233.0
Assam	36	53	78	11.5	12.5	12.2	47.2	116.7
Nagaland	31	68	70	9.9	16.1	11.0	2.9	125.8

state in that year. The last two columns viz 8 & 9, give the percentage growth of per capita consumption over the 10 year period for each state—of 89-90 over 79-80 in column 9 and 89-90 over 85-86 in column 8.

This Table at one glance gives all pertinent information relating to the entire field and absolute as well as relative growth of electricity supply in the various states of the country. Not only the existing status is seen but also the trend of change in the most significant last 10 years.

The position revealed by the table would readily be judged as distrubing. The status existing is not equitable and there is no trend which can work to correct it. We may first note some important aspects as shown by this table.

(i) The per capita consumption of some major states is as small as 1/8 that of the best state.

(ii) Even if the highest state is left out as fortuitous (even though the position is maintained over 10 years) the proportion to the 4 best states is of the order of 1/5. These orders of differences are obviously too large when considered for major states. It may be remarked that there are still further large differentials within the badly off states, but presentation here is limited to state basis, due to lack of data.

(iii) On the basis of relative per capita electricity consumption of 1989-90, the states could be grouped in the following similar levels.

No of States.	*Relative per capita electricity consumption*	*Population covered %(81 census)*	
		Million	*%*
4	55 to 100	126.3	18.8
2	40 to 50	85.3	12.7
3	30 to 40	139.6	20.8
4	25 to 30	61.9	9.26
2	20 to 25	165.3	24,6
5	less than 20	93.2	13.8

Table 2
Total Generating Capacity of Different States Per Million Population

State	*Population Million (1981)*	*Total Installed Capacity* MW*			*Installed Capacity per Million Population MW*			*Installed Capacity Relative Maximum*
		3/80	3/86	3/90	3/80	3/86	3.90	3/90
(1)	(2)	(3)	(4)	(5)	(6)	(7)	(8)	(9)
Punjab	16.7	1593	2400	3800	95.4	143.7	227.5	100
Maharasghtra	62.7	3858	7025	9300	61.5	112.0	148.3	65
Gujarat	34.0	2422	3400	5100	71.2	100.0	150.0	66
Haryana	12.9	1118	1500	2300	86.7	116.8	178.3	78
Tamil Nadu	48.3	2929	3780	4800	60.6	77.8	99.4	44
Karnataka	37.0	1334	2490	3600	36.0	67.3	97.3	43
Andhra Pradesh	53.4	1930	3500	5000	36.1	65.5	93.6	41
Madhya Pradesh	52.1	1528	3140	4000	29.3	60.3	76.8	34
Rajasthan	34.1	1040	1760	2500	30.5	51.6	73.3	32
Orissa	26.3	923	1150	1700	35.1	43.7	64.6	28

(Contd.)

(1)	(2)	(3)	(4)	(5)	(6)	(7)	(8)	(9)
Him. Pradesh	4.2	127	180	320	30.2	42.3	76.2	34
Jammu & Kashmir	6.0	206	280	350	34.3	46.7	58.3	26
Kerala	25.4	1011	1141	2000	39.8	44.9	78.7	35
Uttar Pradesh	110.8	3340	4700	7200	30.1	42.4	64.5	28
West Bengal	54.5	2300	3500	3600	42.2	64.2	66.0	29
Meghalaya	1.3	131	150	155	100.8	115.4	119.2	52
Bihar	69.8	1600	2100	2500	22.9	30.1	35.8	16
Manipur	1.4	15	69	69	10.7	49.2	49.2	22
Assam	19.9	162	420	640	8.1	21.1	32.2	14
Nagaland	0.8	4	21	21	5.0	26.2	26.2	11

*Installed capacity here comprise the capacity owned by the State + approximate allocation of shares from Central Projects and Common Projects.

Considering population wise for the country, thus, nearly half of the people have electricity available at a level of the order of one third to one fourth that for the best placed states. This can be seen manifested in many ways-low proportion of rural electrification, large areas without rural electrification, heavy power cuts and restrictions, reluctance and flight of industries, lack of higher education and employment etc. It is to be mentioned here that urban concentrations like Union Territories of Delhi and Chandigarh and the States of Goa and Pondicherry are not included in this and other tables. These have very high levels of power consumption.

The data presented in Table I thus clearly establishes that there is very alarming imbalance in power development in various parts of the country. It should be cleraly realised that difference in levels of electricity consumption directly represent differences in facilities available for power production as explained earlier.

To what extent efficiency of plant operation is instrumental in differences in power availability?

It is very often mentioned that power plant in certain places in the country is not operated at best efficiency which is meant to indicate that such inefficiency is the cause of shortage of power.

A factual study will show this view to be incorrect and bring out the fact that shortfall is due to very much less generating plant in the relevant state.

Table II on previous page has been compiled to bring out the factual position in this regard. This table, in its columns 3,4 & 5 persents installed generating plant available for power production for each of the states listed in Table I, in 1980, 1986 and 1990. In 'Plant available' is included the plant owned by the state as also approximate share from generating plants in common projects and central projects. The shares have been made on an adhoc approximate basis and the totals can be considered fairly correct.

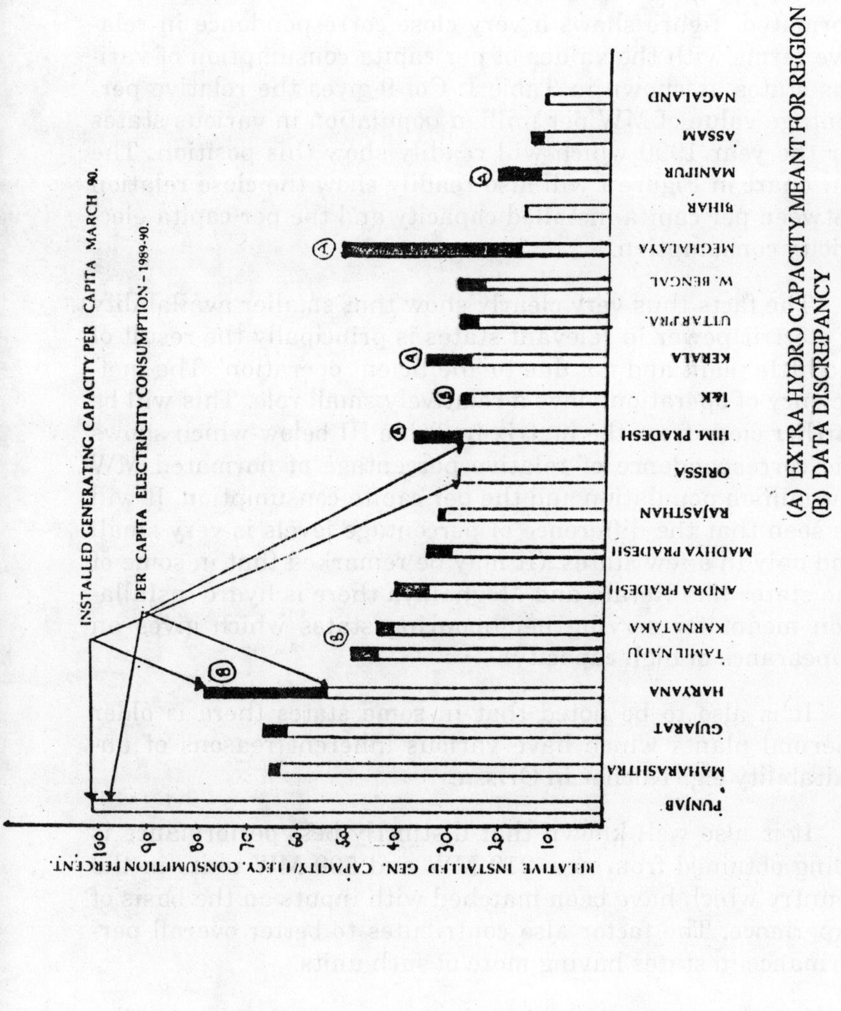

Fig: 1 Relative per capita installed capacity & per capita electricity consumption of states-1989-90
(Actual Values Expressed as percentage of best value for any state.)

Col. 2 in Table II gives the population of the state (based on 1981 census). Columns 6,7,8 then bring out the comparative normated figure-in-terms of MW installed capacity per million population of the state. It would be seen that this normated figure shows a very close correspondence in relative terms with the values of per capita consumption of various states as shown in Table I. Col 9-gives the relative percentage value of MW per million population in various states for the year 1990 which will readily show this position. The bar chart in Figure I will also readily show the close relation between per capita installed capacity and the per capita electricity consumption.

The facts thus very clearly show that smaller availability of electric power in relevant states is principally the result of too little plant and not due to 'inefficient operation'. The inefficiency of operation plays a relatively small role. This will be further clear from the matrix in Table III below-which shows the correspondence of relative percentage of normated MW per million population and the per capita consumption. It will be seen that the difference of percentage levels is very small and only in a few states. (It may be remarked that in some of the states like Kerala and Meghalaya there is hydro installation meant for serving neighbouring states which gives an appearance of high capacity).

It is also to be noted that in some states there is older thermal plants which have various inherent reasons of unsuitability e.g. Talcher in Orissa.

It is also well known that distinctly best performance is being obtained from new 210 MW and 500 MW units in the country which have been matched with inputs on the basis of experience. The factor also contributes to better overall performance in states having more of such units.

It is therefore highly unrealistic to assert that low availability of electricity in various areas is due to inefficient operation. The prime causes is too little installed capacity. As power availability is the infrastructure for virtually everything today-industry, agriculture, employment education and so on-positive measures need to be taken to correct the imbalance. The consequences of imbalance can easily be visualised and need not be elaborated.

Is the Imbalance of Power Facilities Going to Get Corrected by Present Procedures

An elementary consideration of the existing methods of additions to power facilities in the country will bring out an emphatic 'No' as answer.

As a start we may look at data in Table I & Table II forgoing. It will be noted that in no case, in the period of 10 years, there has been any correction of imbalance. On the other hand it has only worsened. This position will be further reinforced if one looks further back in time.

Then, one can look at the basic way of funding of power projects. T:ll now there have been two main categories of power projects viz-

(i) State Projects

(ii) Central Projects.

Besides these, there have been Private Sector Projects but only in a few states. With recent modifications, private sector projects can come up in other places also.

The principal contributor to power generating capacity so far, have been state projects which are funded out of states own financial resources.

Here the axiomatic fact applies that for the states having low electric power capacity, the state GDP itself is small and consequently its total financial resources are also still smaller. Hence, even if it spares the same percentage of its annual financial resources as a better off state or in some cases may be higher, it still remains very meagre. This position will undoubtedly be shown by power plants of different states in every plan. On the other hand, the power deficient states need to take up construction of proportionately more generating capacity to make good their backlog, while there earning

Table 3

Relative Position Showing Relation of Specific Installed Capacity and Capita Consumption

Relative Install Edcap. MW Per Mill ion Popu (% of High Est Value)	Relative Per Capatia Electricity Consumption of States (per Capita Consumption of State Expreised As % of Max. Consumption of Any State) 100-80	80-55	55-40	35-30	30-25	25-20	20-15	<-15
100-80	Punjab							
80-55		Maharashtra, Gujarat, Haryana.						
55-40			Tamilnadu, Andhra, Karnataka				Meghalaya	
35-30				Rajasthan, Madhya Pr.	Himachal, Kerala			
30-25					Orissa J & K	UP W B		
25-20								Manipur
							Bihar	
20-15								Assam
<-15								Nagaland

The related per capita consumption and related MW Per million population are corresponding to each other in all states except 6 states of these, 4 states viz H.P, Kerala, Meghalaya and Manipur have extra hydro capacity for peaking purposes of the region.

from small existing generating capacity is small. This leads to a vicious circle of project delays.

Let us consider now the Central Sector Projects. This is tantamount to a part of divisible financial resources being given to states in kind instead of being given in cash. However, while in the form of financial resources it would be distributed according to particular formulas, when it is changed to distribution in the form of power, the formula gets only 50% weightage. The other 50% weightage is not to deficiency-but to energy consumption. Thus from a central project, an electrically better off state will get proportionately more benefit and the poorly off state will get less due to the adverse alteration of formula.

It is clearly evident that state projects as well as central projects, under the present approach can not correct the imbalances of power capacity.

Besides the financial problem their are a few other surprising factors that are hindering power capacity in particularly power deficient states. One of these is the orchestrated obstruction to some multi purpose (i.e. irrigation-cum-power) river valley projects. While many similar projects are under construction or in operation for decades in other parts of the country without any known adverse effect.

Measures that could be taken to correct imbalance of Power capacity

It would be seen from previous discussion and data that imbalance of power capacity in different parts of the country has reached alarming levels, detrimental to homogeneous development and further that existing approaches of power planning and funding will never be able to correct these.

Without beating about the bush, it has to be clearly recognised that the imbalance can only be corrected by providing addition resources to the deficient states-whether the additionality is in financial terms or in terms of power generation capacity.

This suggestion should not be considered shocking. Such additionalities have been provided to many states in the past since the period of planned development. It will seen for instance that all the states which are very well off in power capacity have recieved such 'additionalities'. The additionalities have taken forms like:-

(i) Additional funds provided for particular reasons.

(ii) Funding as National Projects in the initial years of independence.

(iii) Power projects constructed for exclusive used under some other sector such as mining or atomic power.

(iv) Large private sector projects especially those funded by multilateral agencies.

(v) Funds from outside sources forming additionality to plan resources.

Such additonalities have been greatly helpful in creating or maintaining the lead in power generation in various states.

What is therefore needed is that similar help be extended to other areas which have laggesd so far. The states which have a lead in power will continue to maintain it as nothing can stop it. However a conscious review needs to be made to help those who have fallen behind.

It has now been consciously decided by the nation to involve the private sector in power in a big way. The approach that could be adopted now could be:

(a) allow private sector to have the principal role in addition of power capacity in areas of the country which are well off in power. This will be profitable to the private sector also and attractive for them.

(b) channel bulk of the Public Sector resources to provide additionalities' in the deficient states to bring them up to a reasonable level, comparable with others. The additionalities could be in the manner similar to those which have been followed earlier for different states, whether in cash or in kind.

The above suggestions take full care of the fact that the growth of electricity in advanced areas connot and of course should not be hampered but these places are now self sustaining in growth and will attract it by themselves. At the same time the country will devote attention to bringing other areas which have lagged behind due to defects in the process of planning. The precise way in which corrective efforts should be made can be more exhaustively worked out, but it should be recognised that there is no time to be lost for correcting the appalling inequities in the availability of electricity production capacity on population basis in different parts of the country-which stand at proportions of 1 is to 4, 1 is to 5 or even 1 is to 10 for different places. Immediate action needs to be taken to bring up the position of lagging areas closer to that of more fortunate areas by additional assistance such as has been provided to many places. What needs to be recognised is that assistance requires sufficiently massive additonalities to be given to the states which are poorly off.

Conclusion

In the foregoing paras specific data has been presented to bring out the fact that development of power generation facilities has become highly imbalanced in different parts of the country on a population basis. The inequities are so heavy that power available in various major areas only amounts to 10,15 or 25% relative to other areas in the result of not enough power generation facility having been provided due to inherent defects of the planning process. The often projected reason of poor operation of some old power plants is not the cause of the power deficiency as shown by data in the foregoing-although as will be natural in areas of grave shortage of any facilities some inefficiency and malpractice is not absent.

The alarming imbalance in power facilities calls for immediate corrective action. Since electric power forms today a constituent of all economic activity-industry, agriculture, education, services and so on, the poor status of electricity represents imbalance of all elements and necessary correction brooks no delay.

The main corrective has to come from providing the needed funds or alternatively the corresponding generating capacity to the deficient areas. For simple logical reasons, it will be seen that present process of found allocation can not provide any correction. It is essential that suitable additionality be provided for correction of imbalances in deficient areas-such procediness have in the past helped many areas which are well off today and the need is to extend it to others.

If this is not given the requisite immediate attention the progressively deteriroting imbalance of power will lead to imbalance in all economic spheres with consequent adverse social effects. With the present era of liberalisation, it should easily be possible to make modifications of planning to correct this major source of inequity which has already resulted in many difficulties. The public funding which will have the major role in infrastructure development should be rearranged to provide the essential additionalities to rectify the defect which have been allowed to grow to its present enormous proportions.

CHAPTER 5

Environmental Auditing in the Indian Context

Ms. Manvel Alur,
Research Associate, Tata Energy Research Institute,
Banglore 560 002.

Environmental conservattion has so far been seen as a "luxury" addressed by the richer nations. This is true even for environmental auditing. But the changing attitudes in India have given science - even a new one such as environmental auditing - a chance to prove its potential for a better environment.

The definitions, concept, scope, advantages and methodology of an environmental audit is given to strengthen the readers understanding of it. With the recent initiation of this process in India, the major teething problems, its reception by the industry and the possible tangential developments in the future, are analysed.

Introduction

The growth of the Indian industry has been steady over the past two decades. The continuous rise in consumerism by the vast population has put tremendous pressure on life-sustaining natural resources. The need for environmental conservation is thus warranted, but adapting this "luxury" is affordable in technologically richer countries, whereas its priority for a developing country like India, which looks towards fulfilling its basic requirements of food, shelter and clothing, is understandably low.

The industry has been casting a wary eye at the attempts made by the Indian Government to curb irresponsible attitudes towards environmental issues. Insofar as Legislation is concerned, the Government has only been successful in setting guidelines and standards for pollution levels, but their inadequacy in its enforcement, have given room for industries to take undue advantage and encourage their "...not my problem" attitude, instead of inculcating responsible action. Corruption within the system has only strengthened this indifference further.

The potential of industry to take a leading role towards a sustainable future is undoubtedly there, but only a change from self-centrism to socio-certrism can make this at least possible. Pulling away from an almost impossible situation of attitudinal change within our system of governance, a need for each one taking the responsibility for OUR environment, is the call of the day.

Environmental management is a way by which institutional attitudes could be geared to obtain an efficient and just way of utilising resources and at the same time ensuring economic sustainability.

Environmental auditing is one such management tool, Still new, even in the developed world, its potential, since its inception in the early seventies, has been realised and its future importance to aid in repairing and maintaining an ecological balance is in no doubt.

Environmental auditing essentially entails an increase in the efficiency in utilisation of resources through waste minimisation and technological innovation.

Presently, the concept of environmental auditing in India will only be accepted if the industry sees definite economic benefits in it. In addition to lethargic attitude towards technological upgradation and change, large capital investments will be a major drawback unless adequate incentive is offered. It must be understood and appreciated though that environmental auditing plays a dual role in providing financial benefits from product maximisation and improvement through efficient material utilisation and simultaneously clean up of the environment resulting from reduced waste output. It is a conceptual and technological shift from the "polluter pays" principle to "pollution prevention pays".

This chapter deals with the advantages, overall scope and methodology of environmental auditing and briefly analyses how it is presently viewed and will be received by the Indian industries.

Defining Environmental Audits

The relatively new science of environmental auditing is undergoing constant refining and evaluation of its definition and methodology. But the United Nations Environmental Programme (UNEP) has defined and set some guidelines.

Environmental auditing can be defined as "..a basic management tool comprising a systematic, documented, periodic and objective evaluattion of how well environmental organisation, management systems and equipment are performing". (UNEP, 1988)

Advantage of Environmental Audits

Viewed from both an economic and social angle, environmental auditing has definite advantages to offer.

The advantages of environmental auditing are :-

- it identifies and develops an internal process of waste minimisation through recycling and reuse of resources;
- it encourages the use of low-waste technologies (keeping the goal as "zero waste") and a prudent utilisation of the resources, which is economically attractivee to both the company and the nation as a whole. These financial benefits accure primarily as a result of -
  - decreased material costs;
  - decreased waste management costs;

- improved product quality;
- enhanced productivity;
- decreased worker health risks;
- decreased environmental hazards;
- decreased long term liability for clean up of waste materials; and,
- improved public image for the company.

- it processes the identification of compliance problems and pin-points both strengths and weaknesses.

it encourages continual improvement in production processes.

The need for a systematic and programmed method of evaluating environmental costs due to developmental activity such as industrialisation, justifies the need for environmental auditing as one such method.

Scope of An Environmental Audit

The scope of environmental auditing is largely dependent on several variables - the size and type of the plant and the requirements of the company; but the general scope of environmental audits is wide and it is important to look at it in its entirety.

It covers the complete gamut of organisational structure and management. This would encompass -

- analysis of company management systems in relation to environmental issues;
- analysis of company environmental policy;
- review of environmental training/education;
- review of employment occupational health supports;
- review of technical/analytical support;
- problem solving through information transfer;
- possible areas for improvement in environmental management and policy; and,
- long term planning aspect for environmental sustenance.

Since one of the main aims of an audit is to facilitate management controls of environmental practices, and to enable the company to assess compliance with company and government policies (regulatory requirements), it is essential to assess the legal aspects of environmental management. The legal strength of the company has direct correlation to the extent of environmental damage. To cover this environmental auditing should look at -

- the compliance status;
- previous compliance problems and accidents;
- the existing and required environmental organisation (legal); and,
- the company knowledge of environmental regulations/ laws.

The major thrust area of environmental auditing is in pollution prevention. It identifies and assesses the economic feasibility of potential areas where this could be practised. This scope is vast and the major points that would be covered here are -

- the production process;
- review of the population records;
- review of the existing pollution abatement technologies adopted by the company;
- present emission levels (air/water/soil);
- review of monitoring technologies use by the company;
- study of water and soild waste input-outputs and the potential for recycling and reuse;
- energy audits;
- fuel substitution possibilities;
- feasibility and type of process modifications; and,
- end-of-pipe technologies.

Considering the many advantages and wide scope of environmental auditing, it is imperative that the audit team has expertise in the areas covered under the audit. Presently there does exist a vast sea of knowledge. Clearly, more important is the integration of this for a complete and successful undertaking of environmental auditing to provide strong scientifically backed understanding of the problems and solu-

Types of Environmental Auditing

Preliminary Audit	*Detailed Audit*
● A preliminary audit gives a general idea of the plant operations, while identifying sensitive areas of environmental concern.	● A detailed audit is preceded by a pre-audit. The audit quantifies the environmental costs and also identifies strengths weaknesses of the production system.
● It highlights the pollution aspects (air/water/soil) and the potential for recycling and reuse. Sensitive points are prioritised in this audits.	● An evaluation of the production plan and pollution controls is undertaken. Based on this, options for pollution abatement and regulatory compliance is recommended.
● Depending on the plant size, a preliminary audit is carried out between 2 to 10 days.	● This audit takes between 1 to 10 weeks depending on the size, extent of pollution and resource saving potential.
This audit requires in-death knowledge of the production process and a detailed questionnaire.	● The pre-requisites of a detailed audit in addition to that of the preliminary audit are :- - 1 to 2 pre-audits - schedule and planning for montoring/rcporting/ finalising - monitoring instruments - actual on-site monitoring/measurements questionnaire(s) - support staff

tions within an industry. It must be stressed that the auditors role is NOT that of an environmental police-person.

Depending on the requirements of the company or the government, an environmental audit could be done at various levels. These are :-

- Unit level
- Plant level
- Sectoral level
- Corporate level and
- Regional level

Methodology for carrying out an Environmental Audit

There are several variations in the methodologies adopted for environmental auditing depending on the company and the audit team. One such methodology (guidelines set by UNEP) is illustrated below.

The basic steps of an environmental audit can be summarised to consist of :-

- preparation at the site
- fact-finding, discussion on-site
- evaluation of findings
- assignment of responsibilities and deadlines
- reporting
- follow-up

The main requirements to conduct an environmental audit are:-

- trained auditors
- monitoring and measuring instruments
- audit plan

With this basic idea of the scope, advantages and methodology of environmental auditing, a brief analysis of the way industry would respond and the major teething problems that could be encountered, is discussed below. Some of the points

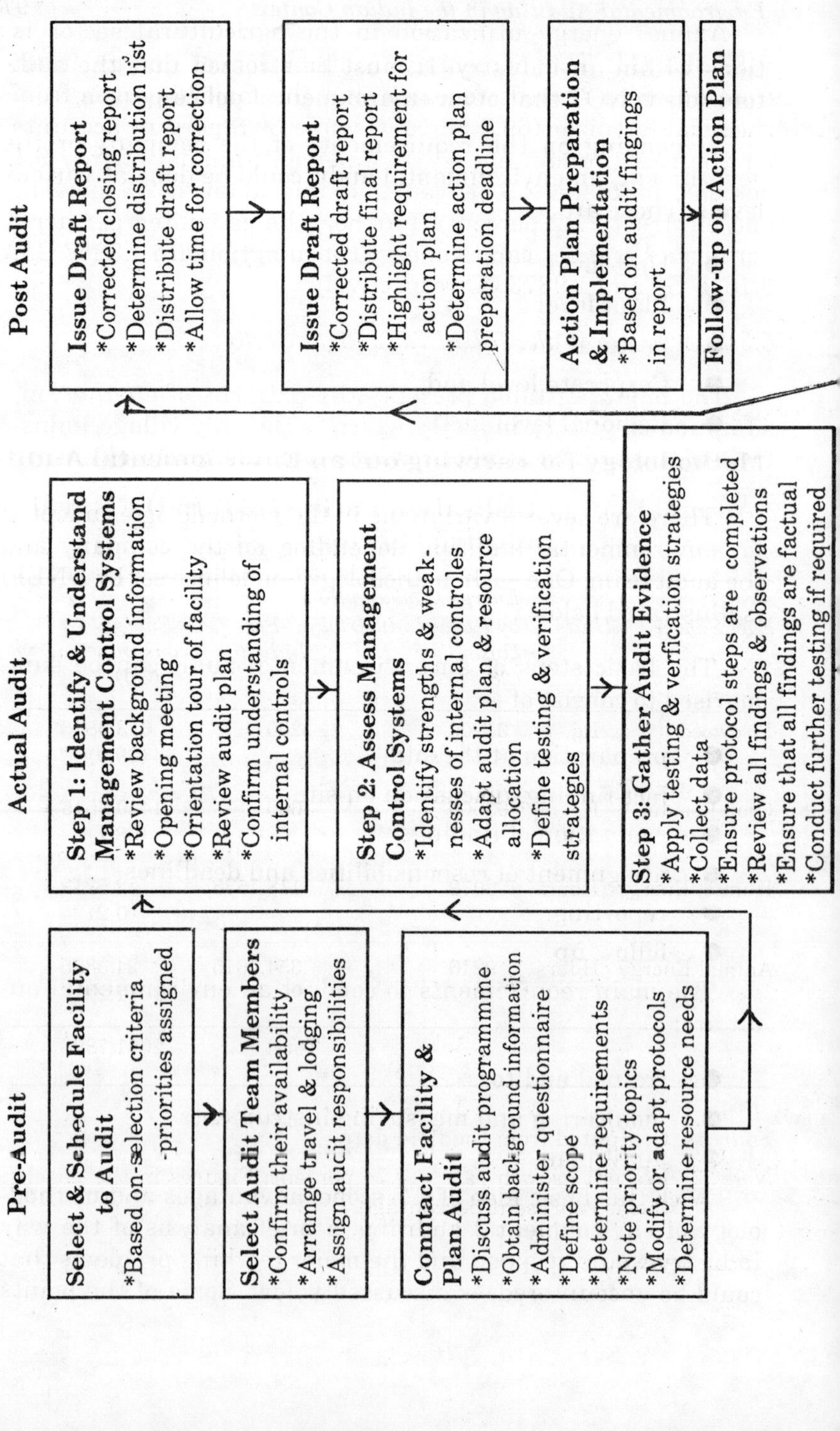
Pre-Audit
Select & Schedule Facility to Audit
*Based on-selection criteria -priorities assigned
Select Audit Team Members
*Confirm their availability
*Arramge travel & lodging
*Assign aucit responsibilities
Conntact Facility & Plan Audit
*Discuss audit programmme
*Obtain background informaion
*Administer questionnaire
*Define scope
*Determine requirements
*Note priority topics
*Modify or adapt protocols
*Determine resource needs
Actual Audit
Step 1: Identify & Understand Management Control Systems
*Review background information
*Opening meeting
*Orientation tour of facility
*Review audit plan
*Confirm understanding of internal controls
Step 2: Assess Management Control Systems
*Identify strengths & weaknesses of internal controles
*Adapt audit plan & resource allocation
*Define testing & verification strategies
Step 3: Gather Audit Evidence
*Apply testing & verfication strategies
*Collect data
*Ensure protocol steps are completed
*Review all findings & observations
*Ensure that all findings are factual
*Conduct further testing if required
Post Audit
Issue Draft Report
*Corrected closing report
*Determine distribution list
*Distribute draft report
*Allow time for correction
Issue Draft Report
*Corrected draft report
*Distribute final report
*Highlight requirement for action plan
*Determine action plan preparation deadline
Action Plan Preparation & Implementation
*Based on audit fingings in report
Follow-up on Action Plan

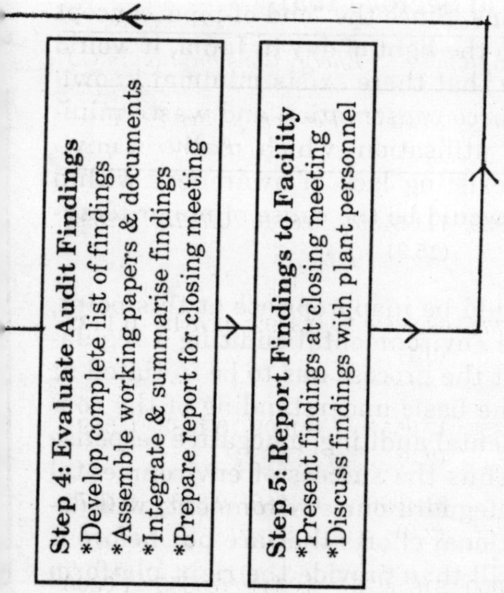
Step 4: Evaluate Audit Findings
*Develop complete list of findings
*Assemble working papers & documents
*Integrate & summarise findings
*Prepare report for closing meeting
Step 5: Report Findings to Facility
*Present findings at closing meeting
*Discuss findings with plant personnel

of this section are drawn from a survey conducted to find out the industrialists view of general environmental issues and environmental auditing in particular.

Response of the Industries

There is no doubt that, only a small percentage of the Indian industries have adopted pollution control technologies to date. There are several reasons for this inaction on their part: ignorance, inadequate law enforcement and irresponsibility are some of the reasons. Since the "end of-pipe' concept has itself only recently seen the light of day in India, it would not be inappropriate to say that there exists minimal knowledge of the concept of resource conservation and waste minimisation through efficient utilisation, which really is environmental auditing. This existing lack of awareness within the industrial community, would be the cause of major teething problems.

Considering this, it would be inappropriate at this point, that the Government make environmental auditing a regulatory requirement. No doubt the process has to be initiated at the earliest, but without the basic understanding of the concept and need for environmental auditing, a negative response would not be surprising. Thus the success of environmental auditing as a process to safeguard our environment, will depend largely on the educational efforts that are put before it is made mandatory; this will then provide the right platform for including environmental responsibility among the industries. Education should be an intensive and continuous process, which would refine and progress in its effectiveness.

Presently there lacks institutional responsibility towards the environment and the only feasible way seems to be to make environmental auditing mandatory. But here again the success would be enforcement dependent. The general psyche of the industries today towards environmental issues, is one that is a "trouble-maker". This attitude needs a desperate transformation.

Another aspect of environmental auditing that would come under consideration sooner or later, is one of authority. Who must be authorised to undertake it? This would decide the extent to which the purpose of environmental auditing is met. If it is a process that is not marred with nepotism and dishonesty it would be, otherwise not. It would be more effective in India, if environmental auditing was initiated by non-governmental institutions which have the credit of higher performance efficiency as compared to governmental institutions. It will of course, not be possible without strong and continued co-operation extended by the government authorities, since enforcement lies only within their jurisdiction.

Now industry feels that environmental auditing is not a priority for India. The relativity of priorities, make it difficult to decide on this. If it is compared with the basic requirements of our people, it would lie on the lowest rung of the priority ladder, but in relation to industrial development and the seriousness of environmental degradation, it would definitely be one of the main priorities.

One of the main objections voiced by industry on introducing environmental auditing, would be that of high initial monetary investments. This could be overcome by a system of tax credits or market incentives. Profits accrued by environmental auditing have to be stressed as this is the only language industries respondy to; but if this became their main contention for non-implementability it would be inexcusable. A point often overlooked, is that of environmental costs which is an important way to conserve our precious resources, and one that everyone has to bear. It would be interesting and an asset for environmental auditing, to know what the pay-back period is and the economic profits they would get.

Future tangential development of environmental auditing, will be consequential to the way in which its impetus is given during the foetal stages. Sustained efforts of the authorities and institutions undertake it will also determine its growth.

Therefore, in conclusion, at this point it would be infantile and wrong to assume that environmental auditing can offer only a systematic and progressive method of evaluating and moderating industrial operations aiming to provide a sustainable form of development, through optimisation and rational use/reuse of resources. It must be accepted as an evolving science drawing no limits to its potential.

REFERENCES

1. Industry and Environment, Vol.11,#4, Oct/Nov/Dec 1988. pp. 1-2
2. Environmental Auditing, Technical Report Series #2 1990. Report of UNEP/Industry and Environment (UNEP/IEO) Workshop. Paris 10-11 January, 1989.
3. Singh, Raunaq, Paper Presented at the Seminar on Environmenatal Management in Indian Industry, 23-24 January, 1992. New Delhi.
4. Ties H. van Kempen, "Achieving economic Benefits Through Effective Environmental Management: examples from the United States", Paper Presented Ibid.
5. Standley, Stacy, "Pollution Control and Environmental Protection - How to make them pay", Paper Presented Ibid.
6. Viswanath B, "Environmental audit: Looking beyond its scope", Fortune India, October 15, 1991. pp 5-9
7. Pearce D W, Environmental Economics, 1976.
8. Karam R K and K Z Morgan, Editors, Energy and the Environment: Cost Benefit Analysis, 1976. pp 479-502.
9. Dryden I G C, General Editor, The Efficient Use of Energy, 1982. pp 489-502.
10. "Environmental auditing is here to stay", Water/Engineering and Management, March 1982.
11. Industry and Environment, Vol.3, July/August/Sep 1984.pp 3-50.
12. Harrison L L, The McGrawhill Environmental Auditing, Handbook.
13. Cahill L B, Environmental Audits, 2nd Edition, 1983.
14. Return S S and C L Sengupta, "Environmental Management the Indian Scenario as Applied to Industry: Opportunities in Technology and Services", 1992.

CHAPTER 6

Energy Scenario of an Indian Village in 2001 A.D.

Dr. S.Ramaswamy,
(Deptt. of Economics)
and
Dr. T.Tamizh Chelvam,
(Centre for Future Studies),
Gandhigram Rural Institute,
Anna Distt., Tamilnadu

A substantial proportion of energy supply, in India, comes from non-commercial sources like firewood, cattle-dung, vegetable wastes and crop residues. However the futurology and management of existing Rural Energy System are found not to be satisfactory. The present chapter, takes up the case study of Kambiliampatti village of Anna district of Tamilnadu for its total energy system including the energy utilisation pattern.

Authors suggest in the chapter an optimal linear reduction form and other useful remark for future improvements.

1. Introduction

Energy has long been a vital element of economic growth, and human progress has progressively been dependent on the use and adequacy of energy. The development of agriculture and industry, the expansion of infrastructural facilities and the transport sector have necessitated ever increasing investments on energy resources development as well as usable forms of energy. In the household sector, the range of nergy dependent consumer durables has greatly expanded.

As such, the importance of energy to the national economy is now more than its direct contribution to the Gross National Product (GNP), as it not only serves as a basic input but also acts as a catalyst for economic development and growth. Energy consumption is dependent on population size, percapita income, life style, industrialization, urbanisation and public energy policies. Many energy economists opined that energy is indispensable for economic development and its consumption is found to increase with the development of the economy. The stage of economic development attained by a country can be identified by its pattern of generation of energy and its consumption. The energy potential of a nation has direct correlation with economic growth and prosperity. The percapita energy consumption of a country is a key indicator of its stage of economic development.

2. Energy Economy of Rural India

India has more than 5.6 lakh villages and 70 per cent of its population is living in villages. It is noted that India's present level of energy consumption is not only low, but substantial proportion of energy supply is accounted for non-commercial sources such as firewood, cattle dung, vegetable wastes and crop residue whose production and consumption data are quite scanty. It is a well known fact that more than 80 per cent of the energy consumed in villages comes from non-commercial sources, which comprise more than 40 percent of the total energy consumed in the country. The burining of cattle dung, wood and other non-commercial fuels in the villages have been a traditional practice because of non-availability of cheaper alternative fuels. Cooking in Indian villages is done almost entirely with firewood[1], dung cakes and agricultural wastes. The natural resources to the tune of 133 million tons of firewood, 73 million tons of cattle dung, 41 million tons of agrowastes and other bio-mass are burnt annually in primitive chulhas at just 12-20 percent thermal efficiency, with huge emission of smoke and grit which are much detrimental to the health particularly women and children[2]. These forms of energy are available at free of cost (zero cost) procurable through personal efforts or at nominal cost.

Besides, women and children spend more time for firewood collection. According to Dr. Anil Agarwal, Director of the Centre for Science and Environment (CSE) in Delhi, "Women in Many regions in India spend 14 to 16 hours working every day. Day after day, year after year, the fuelwood and fodder collection time increases. In many parts, women have literally reached their carrying capacity [3]. The social cost for meeting the rural energy needs has been increasing over the years. A two year study by Ms. Madhura Swamination [4] on energy use pattern, in Garhwal Himalayas Illustrates the rural women collected firewood which meant and average load carriage of 24.5 Kg per trip. In some cases, it went up to 40 Kg. The average time spent on the task was four hours and distance covered was 3.1 Km. Further a study conducted by the Gandhigram Rural University [5] pointed out that the rural women spent 7 hours to collect 29 Kg of firewood from Sirumalai Hills with the distance of 12 Km. Incomplete combustion of such fuels (fuelwood, agricultural wastes, dung cake) leaves a pall of smoke vitiating the atmosphere for every body and adversely affecting the health of the women. The consumers of firewood, unfortunately does not connect the cause and effect between burning firewood and floods.

3. Energy Consuming Sectors In Rural Areas

The main energy consuming sectors in the Indian villages are (i) Household sector, (ii) Agricultural Sector, (iii) Transport sector, (iv) Industrial sector and (v) Commercial sector.

3.1 Energy Use in the Household Sector

Rural household sector uses commercial energy (Electricity) : 87 percent and Kerosene: 13 percent for lighting and non-commercial energy (firewood, dung cake, and crop residues) for cooking. The only sector of village life where commercial energy has completely replaced by non-commercial energy is domestic lighting. Broadly speaking, the percapita energy consumption in the household sector can be taken as an index of the standard of living of the people. It is not only population growth contributing to household sectors energy growth, but also the tendency towards greater affluence among all population groups. The commercial energy used in the household sector show that the share of electricity has increased, and the share of coal and oil has declined. However, it may be noted that electrification of villages does not automatically improve the economic conditions of the people. In 1986-87 about 100 million households (about 80 percent of rural household and 20 percent of urban household), that is, two-thirds (68 percent) of the Country's 142 million households, depended wholly on Kerosene for lighting. Out of 6.6 million tons of Kerosene used in the country in 1986-87, about 2.2 million tons was used for lighting. The average household consumption for lighting is about 2.2 litres per month. Further, the number of unelectrified households in the country is increasing at the rate of about a million a year. This is due to the number of households increasing at the rate of about 2.9 million a year whereas the number of electrified households is increasing only at the rate of 2.1 million. Under these circumstances, the country has been forced to increase Kerosene consumption at the rate of 7.8 percent a year[6]. Till 1987, about 64 percent of the total number of villages have been electrified. But hardly 10-12 percent of the households in these electrified villages use electricity. The main reason for this disturbing factor is the lack of purchasing power of the household. It is estimated that an investment of Rs.500-1800 per household is required to avail itself of electricity. Hence, only the affluent section of the village society can afford electrification and be benefited. According to the Report of the Fuelwood Study Committee of Indian Planning Commission, the share of rural household energy consumption is 19.8 percent from commercial energy and 80.2 percent from non-commercial energy sources. Out of this 80.2 percent non-commercial energy, firewood constitute 42 percent, Agricultural waste 29 percent, animal dung 24 percent and the remaining 5 percent from other sources.

3.2 Energy Use In the Agricultural Sector

Agriculture remains the main stay of life of the Indian village economy as a whole. All agricultural activities are highly dependent upon either in the form of commercial energy or non-commercial energy. Three forms of energy are mainly used in the agricultural sector viz., (a) animate energy which is contributed by man (human energy) and animal (animal energy); (b) inanimate (mechanical) energy which is contributed by machinery (oil or electricity) and (c) chemical energy which is contributed by fertilizer, farm yard manures and pesticides. This sector acts as a producer as well as a consumer of energy. As a producer, it produces different non-commercial form of energy such as agricultural waste, animal waste, rice husk, rice bran, wheat husk, waste from canning and processing units. In agriculture 77 percent of the required energy is derived from animal and human energy. The small component of commercial energy (23 percent) is used in the form of chemical fertilizer (14 percent), electricity (6 percent) and diesel (3 percent). The higher percentage of animal and human energy in Indian agriculture is evident because of an abundant and cheap supply of labour and animal power.

3.3 Energy Use In The Transport Sector

Transportation in villages is based entirely on animal and human power. These modes of transportation are time consuming but economical. The bullocks are used for agriculture work and they are yoked to the carts. Even today bullock cart remains the most important means of transportation of agricultural and non-agricultural products in rural India. At present there are about 150 lakhs bullock carts operating in India, which carry more than 165 million tons of goods per annum. The total investment in this sector comes to Rs.3000 crores and provides direct and indirect employment to 30 million persons. Draught Animal Power (DAP) contribute nearly 70 percent of rural energy input in fields, transportation and irrigation. Bullocks, camels, mules, donkeys, elephant, yak and mule are some of the animals which supply power in different regions of the country. Animal power is a significant resource for a number of applications in agriculture, transportation and other sectors. Eight million work animals provide approximately 40 million hours power for cultivation and transportation. This is equivalent, in energy terms, to nearly the total installed capacity for the generation of electricity in the entire country today. The quantum of freight carried by DAP has been estimated at 3120 million tons KM in the rural areas and 1170 tons Km. in the urban area[7].

3.4 Energy Use In the Industrial Sector

Most of the rural industrial activities such as pottery, brick making, mental work, carpentry, blacksmithy etc., are entirely dependent on traditional and non-commercial sources of energy. No precise data regarding quantum of utilisation of such sources of energy is available. Some of the modern rural industries are using electricity as a fuel for operating the engine.

3.5 Energy Use In The Commercial Sector

Rural commercial sector such as small hotels, petty shops, etc., consumes both commercial and non-commercial forms of energy. No specific statistics are available on the quantity of energy used for rural commercial purposes either for a particular year or for a series of years.

4 Basic Features of Rural Energy System

The Major features of Indian Rural Energy systems are low energy intensity; high domestic energy consumption; heavy dependency on non-commercial energy; fuelwood collection through individual efforts at zero or nero zero private cost; rapid environmental degradation; and fuel in transition from non-commercial free good to commercial commodity.

4.1 Rural Energy Problems

The energy problems in rural areas of Indian economy have assumed crisis dimensions which call for major innovative changes in the institutional mechanism and policy planning. More than 80 percent of the existing energy consumption in the rural areas consists of non-commercial energy which has resulted in large scale disturbance of the environment. A large number of the rural people is too poor to buy commercial energy when supply of those are made available. Living conditions of the rural poor is so low mostly because of the inefficiency of energy use rather than inadequate supply. For example, villagers are spending more time and energy to procure fuelwood than the quantity of energy they got out of if. The chulhas used by the rural households are century old that are open or partly closed and the efficiency is considerably low and thus consumes more fuel and time. Rural energy problem is multi-dimensional and needs to be tackled through multi-dimensional, Integrated Rural Energy Plan.

The existing energy situation in rural areas is part of a vicious circle of low energy input and low productivity and hence rising poverty. The level of energy available in rural areas are insufficient to meet the bare necessities of cooking, minimal farming and even cremating the human corpse.

4.2 Futurology of Rural Energy

When energy supplies are in question all that depends on them is also in question, not only the way of life but life itself. A famine, in the end, is a failure of the energy supply. Because of the economic environment and other constraints, it is becoming difficult to recommend the utilisation of non-renewable energy sources for rural energy. There is a broad realisation that clean fuels from non-conventional sources are the need of the hour. Many biogas experts have described the role of biomass as an alternative source of clean and renewable fuel energy which can greatly replace our future dependence on fossil fuels. They also expressed the view that biogas technology is likely to be useful for solving a number of global problems such as food and energy shortage, desertification and pollution control. Biogas serves as the exclusive source of cooking energy for 20 percent of the 121 million rural households. This has been found be a good substitute for cylinder gas (LPG) which is commonly used in cities. The urgent problems of rural masses such as sanitation, fuel and manure can be solves by installing biogas plants. Many people feel and have expressed their views that the rural household sector may face energy crisis in the form of fuelwood crisis which affects women and children who have to devote more time and efforts to travel longer distances to collect fuel for their daily needs. The fuelwood crisis could be solved through agro forestry, farm forestry, social forestry and other non-conventional, new and renewable energy programmes.

New and renewable energy sources can be grouped into different categories such as (a) solar energy, (b) wind energy, (c) bio-gas, (d) bio-mass and (e) microhydel. Though harnessing solar and wind energy at present appears to be highly cost intensive, with improved technologies and greater impetus, from the Government department to leading research institutions, these sources may in the days to come turn out to be more economical. It is quite interesting to note that our country is supposed to accumulate every year over 300 million tons of agro-residues, out of which hardly a negligible quantity gets utilised as direct fuel. These agro-residues could be converted into solid fuel briquettes, an excellent renewable energy source free from all obnoxious and hazardous gases with very little ash content[8]. The decentralised and renewable sources of energy could meet part of the energy gap in rural areas. But many of the technologies for utilising renewable energy sources are often beyond the reach of the vast majority of the rural population. The role of renewable energy sources in the energy transition even in the next 15-20 years would depend on the extent to which solution to various technical, economic, social and organisational problems can be found. Each of these sources could provide effective substitute for fossil fuels, in the years to come.

The present study analyses the energy-end-use pattern in different sectors of Kambiliampatti village economy. Thus, this study aims:

1. To assess and estimate the level of energy use pattern of the Household, Agriculture, Industrial, Commercial and Transport sectors in the study area.
2. To identify the factors which influence the different sector's demand for energy and.

3. To project and estimate the demand for energy in the different sectors in the study area.

Profiles of The Study Area - Kambiliampatti Village

The area of the present study viz., Kambiliampatti village falls within Sanarapatty Block of Dindigul Anna District of Tamilnadu. According to our survey, the total population of this village is 1005, of which the adult accounted for 63.5 percent and children accounted for 36.5 percent. In this village, 51 percent of the population are literate. Among the literates, male account for 65 percent and female 35 percent only. In order to understand the geographical, demographic, agricultural and economic development profile of the area under study, it is necessary to have a close look on the Kambiliampatti village as a whole.

Kambiliampatti village is bound by Akaraipatti on the west, Kambiliampatti on the east, while in the North and South Kattupatti, Kuthampatti villages are situated. It is situated at a distance of about 20 Kms from Dindigul Anna District headquarters viz Dindigul. The important road link between Dindigul and Senthurai passes through this village.

The occupational distribution of this village indicates that 52 percent of the population are cultivators (farmers) and 48 percent are landless. Further, in this village, the total number of families are 211, of which 87 are agricultural families and 124 are non-agricultural families. Among the landholders, the proportion of marginal farmers are much higher (66 percent) than the small and big farmers 17.5 percent and 16.5 percent respectively. Due to heavy population pressure upon agricultural lands, most of the families in the village hold small size agricultural lands.

The village is composed of various caste segments in which the proportion of backward castes is dominant. In this village there are nine caste groups. As regards households, the Chetty (96 families) are found in the highest proportion (45.4 percent), followed by Reddiar (39 families), Vellalar (23 families), Gounder (18 families) and others (31 families). As regards the average size of the households, 4 persons constituted the average size of the family in the village.

Primary Sources

This study mainly depends on the Primary data. For the purpose of primary data, major techniques applied are : Observation, Informal talks, and Interview schedule.

The survey method was used as a tool for data collection to find out the complete picture of energy consumption pattern of the different sectors (household, agriculture, industrial, commercial and transport) of Kambiliampatti village. For the purpose of present investigation, the census method (complete enumeration) has been thought to be the most suitable method and hence all the households of this village had been interviewed.

The households have been divided into four substrata on the basis of land holding status. These are land less, marginal, small and big farmers. The researchers have met all the households in person at their residence which enable them to obtain more reliable data and information. A very precious and well prepared Interview Schedules were used as tool for data collection. In order to avoid multiplicity of questions and errors, the researchers have conducted pilot study and thus met 30 respondents (5 respondents from each sector). According to the outcome of the pilot study, necessary modifications were made in the interview schedules. The researchers have posed many questions to women respondents about the energy use pattern in the household/domestic sector. Besides, the researchers have also collected information about the energy use pattern in other sectors (agricultural, commercial, industrial and transport) by making use of different interview schedules.

6. Energy Use In The Different Sectors Of Kambiliampatti Village Economy

6.1 Energy Use In the Household Sector

The following table explains annual energy utilization

Annual Energy Utilisation Pattern (In GJ) In the Household Sector - 1991

Energy Sources	*Unit*	*Energy Consumption*			*Energy Consumption in GJ*		
		Total	*Per House Hold*	*Percapita*	*Total*	*Per House Hold*	*Percapita*
Firewood	Kg.	89028	421.93	88.59	1751.9731 (47.6)	8.3031	1.7434
Crop Residue	Kg.	99072	505.46	106.86	1493.33.21 (40.6)	7.6189	1.6110
Dung Cake	Kg.	8892	52.30	11	18.6732 (0.5)	0.1098	0.0231
Kerosene	Lit.	10020	52.18	11	375.7500 (10.2)	1.9568	0.4125
Electricity	Kwh.	8664	47.34	10	39.1904 (1.1)	0.1704	0.0360
				Total	3678.9188	18.1590	3.8260

Source : Computed from the field data.

Note : Present growth rate=3.63 percent; Figure in the parenthesis represent percentage.

pattern of households from different sources. Significantly higher percentage of energy come from firewood (47.6 percent) and insignificant percentage from dung cake (0.5 percent) as well as from electricity (1.1 percent). Crop residue occupy the second place with 40.6 percent and kerosene third place with 10.2 percent. Therefore, it is concluded that dung cake and electricity fall into an insignificant consumption lot when compared to higher percentage use of firewood and crop residue. This may be due to abundant availability of firewood and crop residue in the study area. It is significant to note that kerosene consumption is considerably more when compared to dung cake and electricity. This may be due to the regular supply of kerosene through Public Distribution

System (PDS). The estimated total energy consumption of different forms of energy is 3678.9788 GJ of which, 12 percent of energy come from commercial sources and 88 percent of energy come from non-commercial sources.

6.2 Energy use In The Agricultural Sector

The table given below explains the annual energy u⁺iliza-tion pattern of agricultural sector from different energy sources.

Annual Energy Utilisation Pattern (GJ) In The Agricultural Sector - 1991

Energy Sources	*Unit*	*Energy Consumption*		*Energy Consumption in GJ*			
		Total	*Per acre*	*Total*	*Per acre*	*Per Farmer*	*Per capita*
Human Energy	Hours	137208	802.3859	114.8430 (15.3)	0.6716	0.7575	0.2392
Animal Energy	Hours	25800	150.8771	68.2668 (9.1)	0.3992	0.7848	0.1422
Organic Manure	Kg.	25810	150.9356	17.2552 (2.3)	0.1009	0.1983	0.359
Pesticide	Litre	2121,2498		21.4809 (2.9)	0.1256	0.2469	0.4475
Fertilizer	Kg.	17100	100.0000	518.8276 (69.2)	3.0341	5.9635	1.0808
Electricity	Kwh.	2435	14.2397	8.7660 (1.2)	0.0513	0.2306	0.0417
		Total		749.4395	4.3867	8.1816	1.9873

Source : Computed from the field data.

Note : Present growth rate = 3.1 percent; Figuires in the parenthesis represent percentage.

Annual energy utilization in the agricultural sector is 749.4395 GJ. The percentage contribution of energy input from different sources revealed that the major input is from chemical fertilizer (69.2 percent) while 30.8 percent accounts for the rest (human energy 15.3 percent; animal energy 9.1 percent; pesticide 2.9 percent; organic manure 2.3 percent and electricity 1.2 percent). Further, the estimated per acre, per farmer and per capita energy consumption are 4.3867 GJ. 8.1816 GJ and 1.9873 GJ respectively.

6.3 Energy use in the Rural Industrial Sector

The following table presents the patterns of energy consumption of pottery industry which is the only village industry existing in the study area. The various forms of energy

Annual Energy Utilisation Pattern (6J) In the Rurul Industrial Sector - 1991

Fuel Source	*Units*	*Energy Consumption by Rural Industrial Units (26)*	*Per Industry*	*Energy consumption (GJ)*
Firewood	Kg.	31200	1200	613.9816 (29.9)
Crop Residue	Kg.	93600	3600	1410.8515 (68.7)
Human Energy	Hours	3072	118.1538	2.5712 (0.2)
Animal Energy	Hours	9216	354.4615	24.3855 (1.2)
			Total	2051.7898

Source : Computed from the field data.

Note : Present growth rate = 4.74 percent; Figures in the parenthesis represent percentage.

used for pot making are Human energy, Animal energy, Crop residue and Firewood. Animal energy is used to bring soil from far away places; collect crop residue and firewood from nearby gardens and forest area. Animal energy is mostly used to collect soil and occasionally used for collecting firewood and crop residue. Both crop residue and firewood are used for kilning the wet pot.

In the study area, large quantity of crop residue is available and naturally the pot makers are using crop residue for kilning the wet pot. In this village, there are about 26 pot makers who are engaged in pot making. The total energy consumed by pottery industry is 2051.7898 GJ per annum, of which 0.2 per cent of energy supplied by Human energy, 1.2 per cent of energy supplied by animal energy, 68.7 per cent of energy supplied by crop residue and 29.9 per cent energy supplied by firewood. It is noted that most of the energies are available at free of cost or zero private cost. Further it is observed that only 26 families are engaged in pottery making and 77.87 GJ of energy is used by a potter per annum.

6.4 Energy use in the Commercial Sector

The annual energy use pattern in the commercial sector is presented in the following table. Major components of commercial sector in this village are hotel, tea shop, petty shop, cycle shop, tailoring shop, barber shop, educational institutions, government offices and radio/TV centre. The total energy consumption of the commercial sector is 2864.3979 GJ per annum, of which 47.6 per cent of energy supplied in the form of firewood; 51.3 per cent of energy supplied in the form of kerosene and 1.1 per cent of energy supplied in the form of electricity. In the study area there are 15 tea shops and 9 hotels (major users) which consume 2168.9824 GJ energy per annum.

The major sources of energy for hotel is Kerosene (67.22 percent) which is followed by firewood (32.66 per cent) and Electricity (0.1 per cent). Petty shops normally deal with groceries and provision items. In this village there are about 8 petty shops which consume 3.402 GJ of energy, of which

Table 3

Annual Energy Utilisation Pattern (GJ) in the Commercial Sector-1991

Energy Soruces	*Actual Units*	*Hotel*	*Pettishop*	*Cycle Repair*	*Cloth / Tailor*	*Barber*	*Educational Institution*	*Govt. Offices*	*Total*
Firewood	69,2999 tons	708,4404 (32.66)	-	-	-	-	655.3073 (98.50)	-	1363.7477 (47.6)
Kerosene	39202 litres	458 (67.22)	2.25 (66.14)	2.7 (74.23)	-	4.5 (83.9)	2.6250 (0.4)	-	1470.0750 (51.3)
Electri-city	8492 Kwh	2.542 (0.12)	1.152 (33.86)	0.9372 (25.77)	6.48 (100)	0.864 (16.1)	7.3440 (1.1)	11.2520	30.5712
Total		2168.9824	3.402	3.6372	6.48	5.364	665.2763	11.2520	2864.3939

Source : Computed from the field data.

Note : Present growth rate = 5.82 per cent; Figures in the parenthesis represent percentage.

66.14 per cent of energy comes from Kerosene and 33.86 per cent of energy comes from electricity.

There are about 11 cloth and tailoring shops that consumed 6.48 GJ of energy, supplied exclusively by electricity. There are 2 barber shops, which consumed 5.364 GJ of energy which consists of Kerosene and Electricity. The Educational institutions (Primary school, Balwadi, High School, Chief Minister Noon meal centre) consumed 665.2763 GJ of energy per annum. The educational institutions particularly Noon meal centres consumed a major proportion of energy in the form of firewood, which finally may lead to deforestration in the nearby forest area.

In order to mitigate this situtation, it is suggested to introduce solar cooker in the noon meal centre, which may considerably save firewood and thus reduce the intensity of deforestration.

6.5 Energy use in the Transport Sector

The following table exhibits the annual energy utilization pattern of the transport sector in the study area. The different modes of transport available in this village are Cycle, Bullock cart, Tractor, Scooter and Moped. In this village 80

Annual Energy Utilisation Pattern (GJ) in Transport Sector-1991

Energy Source	*Actual Units*	*Cycle (80)*	*Bullock cart (10)*	*Scooter / Moped(2)*	*Tractor (6)*	*Total (in GJ)*
Human Energy	48710 Hours	34.1412 (100)	6.0264 (13.7)	0.6026 (0.6)	-	40.7702 (20.5)
Animal Energy	14400 Hours	-	38.1024 (86.3)	-	-	38.1024 (19.2)
Diesel& Petrol	3360 Litres	-	-	102.5280 (99.4)	17.0880 (100)	119.6160 (60.3)
Total		34.1412	44.1288	103.1306	17.0880	198.4886

Source : Computed from the field data.

Note : Present growth rate = 4.91 per cent; Figures in the parenthesis represent percentage.

cycles, 10 Bullock carts, One tractor, and 6 two wheelers (Scooters and Mopeds) are available. Besides, people are also using the public transport system. Cycle is the cheapest mode of transport, followed by Bullock cart. The other mechanised modes of transportation are available through tractor and two wheelers. The major sources of energy utilized by the transport sector are human energy, animal energy, mechanical energy through Diesel and Petrol.

It is a well known fact that human energy is exclusively used by cycles which consume an estimated amount of 34.1412 GJ of energy per annum and the estimated human energy used for operation of cycle is 0.4268 GJ per annum. Animal energy is a predominant rural energy source and used in farming and transportation. In rural area bullock cart is operated with the help of bullocks and human energy. Thus, to operate bullock cart, there is a need for both energies namely Animal energy and Human energy.

In the study area, there are about 10 bullock carts which totally consume 44.1288 GJ of energy per annuam, of which 86.3 per cent of energy offered by Bullocks and the rest (13.7 per cent) by human beings. The estimated quantity of energy (inclusive of animal and human energy) used by a bullock cart is 4.4188 GJ per annum. The other mode of transportation available in this village is tractor, which consumes 103.1306 GJ of energy per annum, (in the form of Diesel and Human energy) of which 99.4 per cent of the energy is in the form of Diesel and the rest (0.6 per cent) is in the form of human energy. It is noted that only 6 two-wheelers are available and used by selected households in the study area. The two wheelers consume 17.088 GJ of energy (in the form of Petrol) per annum and 2.848 GJ per two wheeler per annum.

Thus it is concluded that in the study area, the form of energies such as Animal energy, Human energy, Diesel, Petrol are mainly used for transportation. Further the total energy used for transportation is 198.4886 GJ per annum. In future, there is a scope for rising the consumption level of both diesel and petrol, the so called non-renewable form of energy, even in the rural area where people introduce the modernisation through mechanisation.

7. Energy Scenario of Kambiliampatti Village

Forecasting the future demand for energy has always been a hazardous task. The demand for energy in different forms is dependent on a number of factors like the rate and pattern of economic activities, the technologies adopted for production of different goods land services, the efficiencies of use of energy in the different sectors and the extent of substitution of animal power and non-commercial fuels by the commercial fuels, the behaviour of which is difficult to predict on a long term basis. The recent realisation of the severe constraints on the availability of finite fossil fuel resources and the producer's response to such realisation have made it almost impossible to make a realistic long term forecast of the energy demand. In spite of uncertainties in regard to the factors that determine demand, there is need to make long term forecasts of energy demand, as the long gestation of energy supplying projects necessitate decisions to be taken far ahead of the demand for installing production facilities for different forms of energy. In this section researchers forecast the energy demand in various sectors of the study area for the next 10 years. Since the researchers have analyzed the existing usage pattern in energy in various sectors in terms of GJ consumption, the forecast is also in terms of GJ units. The researchers use the following model for the purpose of energy forecast in each and every sector.

Linear Reduction Pattern

N = Number of years of forecast.

A = (Z - Y)/N , where

Z is the proposed growth rate at the Nth year

Y is the present growth rate

The rate of growth R(I) is defined by

$$R(I) = Y + A*I , I = 1,2,...., N$$

The forecast for the ith is defined by

$$F(I) = (1 + R(I))*F(I-1)$$

$$F(O) = \text{Base year data.}$$

At the macro level the Report of the Working Group on Energy Policy (1979) has proposed the growth rate of Agriculture, Industries, Transport and other sectors of economy for the year 1992-2000 AD as 3.93, 7.23, 6.72 and 6.74 respectively. Since the study area is a Micro Level Unit, the researchers assume the growth rates of the Agriculture, Industrial, Transport, Commercial and Household sectors of the village as 4.00, 6.5, 6.1, 6.2 and 5.1 respectively in view of the present growth rates. Based on the above growth rates energy forecast of different sectors of the Kambiliampatti village has been made.

The following table gives the forecast of demand for energy in various sectors in terms of GJ units. Actually the forecast reveals that the energy demand in the Household sector will be almost three times of the existing one, the demand in the Agricultural sector will be more than two times of the present consumption. Similarly the demand in Rural Industries in 2001 AD is 6645.7100 GJ which is nearly three and half times of the present consumption viz 1958.9360 GJ. In the case of commercial and transport sectors the demand for energy in 2001 AD will be 5898.8470 GJ and 521.4096 GJ respectively.

Forecast Energy Demand in Different Sectors of Kambiliampatti Economy (in GJ)

Year	*Household (5.1%)*	*Agricultural (4.0%)*	*Industrial (6.5%)*	*Commercial (6.2%)*	*Transport (6.1%)*
1990	3550.8680	726.9055	1958.9360	2706.8560	189.2069
1991	3678.9188	749.4398	2051.7898	2864.3939	198.4886
1992	4004.0450	796.4563	2277.2540	3069.3150	216.5817
1993	4296.5410	839.0668	2485.3950	3277.1080	233.6591
1994	4641.9830	887.7329	2734.4300	3505.1940	253.4733
1995	5049.3170	943.2164	3032.4820	3755.8150	276.4759
1996	5529.5060	1006.4120	3389.7060	4031.4910	303.2110
1997	6096.0030	1078.3710	3818.8400	4335.0620	334.3353
1998	6765.3440	1160.3280	4335.9070	4669.7280	370.6439
1999	7557.9040	1253.7350	4961.1410	5039.1030	413.1008
2000	8498.8610	1360.3030	5720.1900	5447.2690	462.8791
2001	9619.4360	1482.0510	6645.7100	5898.8470	521.4096

Figures in the parenthesis represent the growth rate.

8. Suggestions

8.1 General

In order to reduce, the fuelwood consumption and conserve the forest near by the study area, the following measures could be brought out immediately.

1. The Government must come forward to supply smokeless chulahs to the households at nominal cost or free of cost to economically and socially weaker section of the community.
2. Energy end use training could be imparted to the rural women so as to conserve the fuelwood through adoption of smokeless chulahs.
3. Community solar cooker could be introduced in the village which will considerably save the forest near by the village.
4. The waste land in the study area could be brought under several foresty programmes (Agro-foresty, Social foresty, Farm foresty, etc.,) which would increase the supply of fuelwood to the human beings and fodder to the cattle.
5. Fast growing fuelwood species (trees) could be indentified and planted in the waste land of the study area.
6. The financial institutions should come forward to provide credit facilities for the purchase of milch animal and cattle under various Rural Development Programmes that enable the beneficiaries to have more quantity of cattle dung. This helps them to introduce biogas plants, that will meet out the fuel requirements of the households.
7. Village Energy Cell and Energy Development committee could be formed at the village Panchayat level, that will motivate the villagers to save energy and environment.

8. Fuelwood could be supplied through Public distribution System (PDS) that will reduce the illegal fuelwood cutting from the near be forest.
9. Rural energy Programme could be integrated with the on going Rural Development Programmes so as to strengthen the Rural Energy System.
10. Micro Level Energy Information System could be developed to disseminate the importance of energy and environment through the national network programmes in TV., Radio and other mass media.

8.2 Agricultural Sector

1. Integrated Fertilizers and Pest Management system could be introduced in the agricultural sector, which will considerably save fertilizer and pesticides.
2. Human and Animal Energy can be effectively utilized by imparting training to the labourers and improving agricultural technology. Thus in turn reduce the quantum of animal and human energy.
3. Modern ploughing equipments and modernised bullock carts could be used, which will enhance the efficiency of bullock power.

As agriculturist can save 80 per cent of diesel and considerable quantity of electricity in his diesel/electric water pump using efficient footvalues, proper size pipes of suitable material, ISI mark pump, prime mover of matching size, etc.

8.3 Rural Industrial Sector

1. Pot making is the only Rural industrial activity of this village which normally use twings, branches and some quantum of firewood. It is observed that there is a much scope for using the crop residue that are available in plenty. The pot makers of this village must use modern kiln, rather than traditional one, so as to save the trees near by forest.

2. Pot making industry requires more Human and Animal energy for collecting firewood and clay. Hence, it is suggested to establish the kiln, where these raw materials (firewood, crop residue clay etc.) are available, that considerably requires less human and animal energy.
3. The pot makers should not be allowed to use the commercial wood and advise them to use the local firewood such as acacia for burning the wet pot.
4. The Rural Industries Department can impart innovative training to the pot makers that enable them to adopt the modern kiln.

8.4 Commercial Sector

1. Solar water heaters and solar cookers could be introduced in the commercial establishments viz hotel and coffee stall, service centres viz, Chief Minister Noon Meal Centre and Backward class Hostel. The introduction of solar cookers and solar water heaters can considerably save fuelwood.
2. Hotel owners are advised to use fuel efficient ISI marked stove that can be considerably save firewood and charcoal.

8.5 Transport Sector

1. Pedal power (Bicycle) is much relevant to Indian villages in general and the study area in particular. Bicycle is an appropriate, ideal, non-polluting and non-expensive technology and sometimes, even quicker method of transport. Thus, to promote pedal power, the financial institutions should come forward to provide financial assistance to purchase bicycles in the study area.
2. The efficiency of Animal Drawn Vehicle (ADV) could be improved by fitting ball bearing wheel, rubber tyre wheel in the cart. This increases carrying capacity as well as reduces the burden and drudgery of the animals.

To save diesel and petrol in the case of vehicles, one can drive a vehicle at a uniform speed of 40-50 Kmp, avoid frequent acceleration, braking, anticipate stops, stops engine at long stops, keep proper air pressure in tyres, get vehicle tuned periodically and thus save at least 10 per cent of fuel.

9. Conclusion

Energy is one of the primary requisite of life. The demand for energy in the recent past has been tremendously growing due to rise in population, on the one hand and modernisation of agricultural, Industrial and the household activities in the rural and urban areas on the other. However the supply of energy has not been growing in the same proportion of the demand for energy. This is turn, creates a disequilibrium in the demand for and supply of energy. To solve the problem of disequilibrium, a micro level blue print should be formulated either to increase the supply of energy or to reduce the demand for energy. The reduction of demand for energy should not affect the economic development at any cost. Adoption of careful and more efficient use of various forms of energy is the viable and feasible solution to the growing needs of energy in rural India.

REFERENCES

1. Smil and Knowland, G. Willian (Eds), "Energy in the Developing Third World", Oxford University Prcss, 1980 p.7; and Batta, C.L., "Forest Future", *Intensive Agriculture* Vol.22 Nos 3-4 , May-June 1984 , p.18.
2. Misra R.C. : "A Consumer's Guide to Cooking Appliances, Cookware and Fuels", *Urja,* Februrary 1986, p.95.
3. Chatterji, Soma A: "Environement Vital for Women" *The Hindu* 13 May, 1990, p. 18.
4. Ibid
5. Narayanan N and Ramaswamy S : Fuelwood Cutters of Sirumalai Hills - A Case Study (Gandhigram Rural University, Gandhigram, May 1991) pp.32-34.
6. Reddy A.K.N. : "Energy Planning - Time for Tough Decisions" *The Hindu,* 23 November, 1990, p.8.
7. Indian Institute of Management: India's Animal Drawn Vehicles (11M, Bangalore, 1984) p. IX.
8. Rangaswamy V.S. : "Bio-mass - A Potential Energy Source" *The Hindu* 14 May 1991, p. 17.

CHAPTER 7

Wind Energy : The New Energy Option For 21st Century

Dr.(Mrs) Dhulasi Birundha Varadarajan
Reader in Economics,
Madurai Kamaraj University
and
A.Napoleon
Research Scolar,
Madras University

Among the various types of renewable energy sources wind energy is stated to be the only proven alternative in the energy structure. Along with the evolution of man from primitive stage to the present civilisation, the wind energy has also moved with time from its ancient period to the present age of sophistication, that is adaptible to various usages like water pumping and power generation. The authors present here the wind energy projects at Kayathar, Tamil Nadu. The chapter covers key aspects like wind utilisation in India, technical aspects of wind power generation, physical process of power generation through wind mills and other economic considerations.

Introduction

In the recent years we have been encountering what is termed as "Energy Crisis". Energy crisis is not only a threat to any economy but also posses a challenge in correcting the deficiency. The energy requirement of any nation is directly proporational to its industrial growth and population. India which is going in for massive industrialisation is a population giant and has to reckon with the very relevant issue of energy crisis.

It is only in recent years that price hike and shortage of oil is being experienced. Fossil Fuels are non renewables and once exhausted can not be replenished. It is this very fear which turns our attention towards alternative sources of energy. Though it has been will over decades since thoughtful people first began to think seriously about a world without abundant supplies of low cost energy. In this regard the Indian Government has laid a great deal of emphasis on the development and harnessing of alternative sourcess of energy.

The alternative energy sources include solar photovoltaics, Geo-thermal power, Tidal power, Biomass energy, solar and wind-power. Among the various types of renewable energy sources wind energy is presently the only proven alternative in the energy structure. Along with the evolution of man from primitive stage to the present civilisation, the wind energy has also moved with time from its ancient period to the present stage of sophistication which can be adopted to various needs like water pumping and power generation.

The commercial exploitation of any innovation is to be considered in conjunction with the trios of

1. Potential availability
2. Practical applicability and
3. Economic viability

Wind power generation is one type of energy alternative which can satisfy the above.

2.1 Wind Utilisation in India

The Indian sub-continent is considered to be a high wind potential zone in Asia which is estimated at about 20,000MW unlike other renewable energy systems, the wind energy involves the simplest form of technology. Wind Energy in Developing Countries prepared in 1988 for the U.S.Department of Energy and the World Bank in which the authors have screened more than 100 developing countries. In applying a comprehensive list of criteria ranging from geographical and meterological aspects from grid capacity and cost competitiveness to technical institution and financial factors they arrive at the ranking where India occupies the second rank having the scope of 79. In our country Gujarat is the most windy state and Tamil Nadu stands next to Gujarat in creating wind energy. In Gujarat in particular the costal strech between Mandvi and Kaudia of about 100 Kms has an estimated wind potential of 5,000MW.

Along the Western Coast we have windy sites in Maharashtra, Deograh which have mean annual wind speed of about 19 to 20 KMPH. On the Eastern coast especially in Orissa, Puri wind potential annually per unit swept area is around 500 kwhr/m^2 yr.

In Tamil Nadu sites like Tuticorin, Coimbatore and Kanyakumari has the wind potential of the order of 520, 650, 610 kwhr/ m^2 yr. respectively. Also we have windy sites in Madya Pradesh (Indore) Jammu Kashmir (Baniha, Tojilla) Uttar Pradesh (Kumaon hills) and North Western India (Himalayan range). All the above sites accounts the minimum wind speed required to establish wind farm for power generation. But wind farms are now popular only in Gujarat and in Tamil Nadu.

In Tamil Nadu wind power areas are on the Western side affluenced by Western Ghat. In order to have reliable and useful data of wind potential for two years or more with hourly averages 20 wind mointoring stations have been installed with assistance from the Government of India, Department of Non-conventional Energy Sources (DNES) is being implemented in Tamil Nadu. Along with the wind monitoring station, Tamil Nadu is having 30 wind mapping stations at different places. Under this programme a cup counter anemometer fined at the top of a 5 metre mast gives the speed of the wind at that height. Readings are taken at synoptic hours viz., at 5.30hrs, 8.30 hrs, 14.30 hrs, 17.30 hrs and 20.30 hrs daily and the average speed of the wind at the location is arrived.

3.1 Technical aspects of wind power generation

Wind Energy is the kinetic energy associated with movement of large masses of air resulting from the differential heating of the atmosphere by sun. Again it is defined as wind is air set in motion by the uneven heating of earth by sun. It is the movement of the air around the earth blat averages temperature extreemes, and produces the surface winds so useful for power generation.

Power in the wind:-

The power in the wind increased with the cube of wind speed. The theoretical wind power is determined by the formula power density.

$$K_E \text{ X } \tfrac{1}{2} \text{P X } V^{-3} \text{ (W/m}^2\text{)}$$

where

P = density of air (1.2 kg/M^3, 20°c, sea level)

V$^-$ = average wind speed in (m/sec)

W/m^2 = Watts per square meter

K = Energy pattern factor,

K_E	Type of wind regime
.6	Polar
2.70	Continental, irregular
1.92 1.54	Costal, maritine.
1.36 1.22	Trade winds

In the case of water pumping wind mills, a reasonable approximation of the effective power output to lift water can be estimated by

$$0.1 \times V^{-3} \ (W/M^2)$$

Wind turbines generally are more efficient and their effective power output to generate electricity can be estimated by

$$0.2 \times V^3 \ (M/M^2)$$

Another formula to calculate power in the wind is given by

wind power (p) = $\frac{1}{2} dAV^3$ (W)

d - air density (kg/m^3)

A - Area swept by the blades (m^2)

V - Wind velocity (m/s)

Physical process of power generation through Wind mills:-

Wind mills erected after assessing the wind frequency and soil conditions of a particular place. Wind mill has three major parts. The first one is the blades which rotates according to the wind speed. The wind Electric generators requires a minimum wind speed of 11 to 13 kms per hour to start generators and they produce the maximum output at about 35 kms per hour.

The second major part is the rotor. In rotor the wind energy is converted into electric energy and supplied through power lines. The blades are attached to a Hub which is designed to rotat at a constant speed of 40 Rpm. This is directly connected to a gear Bose which converts this 40 Rpm into 1800 Rpm. By means of a series of 'V' belts and it directly connectly to the alternator. The alternator is synchronized with the utility grid.

The third main component in a wind mill is the grid connection. Grid system carries the electrical power generated by the rotors. The generation voltage of the wind mill is 400 V. This is connected to a step up transformer 400 V/11KV through a cable of adequate size. One transformer is normally errected for three or four machines. The cables are terminated in a L.V. panel where MC.C.Bs, are provided for each wind mill. The HV side of the transformer is connected to existing 11 KV lines/ 22 KV lines. Upto 4MW separate 11 LV lines can be laid from the sub station. Above 4 MW, it is desirable to errect a 33 KV /66KV / 110 KV substation at the wind farm. Tamil Nadu Electricity Board has erected a 66/ 11KV substation at Kayathar and a 110/11KV substation at Muppandal.

3.3.4. Grid para meters:-

The various electrical parameters insisted upon the wind mill suppliers are as below

a.	Voltage variation	- 400 ± 13%
b.	Frequency Varriation	- 50 HZ ± 2HZ
c.	Voltage and current asymmetry	- 12.5%
d.	Power factor	- Above 0.8 Kg.
e.	Starting current	- Should not exceed full load current.

The space required for erecting single wind mill generator will be only of the order of 10m × 10m. However, if a group of windmills are to be erected in the same location then the distance between the two wind mills coming in a row will have to be 5 times the diameter of the rotor and the distance between the successive rows have to be seven times thee diameter of the rotor.

4. Types of wind energy system

There are two basic approaches for converting wind power to useful energy. The first one is vertical axis wind turbines (VAWT) in which the axis of rotation is perpendicular and horizontal axis wind turbines (HAWT) whose axis of rotation parallel to the wind stream and the ground. Wind power however is normally being used for mainly 2 purposes. One for water pumping and other for electric generator.

5. Economic aspects of wind energy

The following are the main criteria of wind energy regarding economic aspects.

5.1 Wind energy requires high capital cost, but free fuel and relatively smaller percentage of operation and maintenance cost unlike fossil fuel power stations which needs relatively lower capital investment but have recurringg fuel cost which estimates around 10 to 15% and relatively higher operations and maintenance cost.

Subsidies and tax rebates offered to wind energy projects from Government of India and the environmental cost to the society.

The most important feature of wind farm is the shorter generation period.

5.2 Benefits from wind electric generates

The benefits from wind electric generators when estimated interms of avoided costs include.

1. The savings in capital cost of a diesel generators.
2. Savings in the cost of diesel oil used in electricity generation.
3. Savings in capital cost of coal based plant and
4. Savings in fuel costs and operating and maintenance charges.

5.3 The cost benefit analysis of two wind farms in Kayathar, Tamil Nadu had revealed the cost effectiveness of wind energy. The net-present value method is used to find the net social benefit. The net present value is calculated through the following formula.

$$NPV = (Pt\ Qt) \sum_{t=1}^{T} (1+r)^{t} - C_t \sum_{t=1}^{T} (1+r)^{-t} + St + Tx \qquad \ldots 1.$$

$t =$ 1,2, ...T

$P_t =$ Price level at time period 't'

$Q_t =$ Total quantity that can be produced at time 't'

$C_t =$ Total capital cost, including operation of maintenance, depreciation, and other expenses at time period 't'

$S_t =$ Subsidies granted by government at time period 't'

$T_t =$ Tax benefit that is available to that project

$(1+r)^{-t} =$ The discount factor where r=12% (Bank rate).

On the basis of positive results obtained through net present value method for Kayathar wind farms I and II the economic viability and technical feasibility is ensured. [*See Table 1*]

Economic Aspects of Wind Farm in Kayathar:

1.	Installation	Plant I 1988	Plant II 1990
2.	Total capacity	2550kw	6000 kw
3.	Plant load factor	30%	30%
4.	Capacity available/year	765 kw	1800 kw
5.	Price level (Rs.)	1.40	1.65
6.	Expected revenue	93,81,960	2,60,17,200
7.	Capital costs (Rs.)	510 lakh	1200 lakh
8.	Tax benefit (Rs.)	1,91,25,000	4,50,00,000
9.	Asset value (Rs.)	3825 lakh	900 lakh
10.	Depreciation (Rs.)	17.21 lakh	40.50 lakh
11.	Maintenance (Rs.)	12.75 lakh	30.00,000
12.	Net present Value (Rs.)	35.95 lakh	175.6 lakh

Source:- Survey data of the results of the cost. Benefit anlysis.

6. Conclusion

So far wind farms are operating in large scale in the government sector. There is a need for bringing private sector in generating power through wind. If the private sector comes forward to install wind farms then energy crisis may be reduced considerably. The government at present has given permission to private wind mill generators to import under open general license. But this involves high capital cost. So the Government should come forward to start producing wind mills in our country. No customs duty, Excise duty and sales-tax is applicable now. The Indian Renewable Energy Development Agency is extending soft loans at 8.5% interest for the wind farm projects.

To conclude wind energy has been one of the few forms of renewable energy which traditionally has been lost effective and practical during previous eras. The present question is whether the combination of improved knowledge of the wind and wind system combined with the rising cost and economic penalities associated with fossil fuels can again lead to the wide, scale contribution of energy from wind. Recent experiments and assessments lead to the probability that this will occur.

REFERENCES

1. Carlman, I., *European Wind Energy Conference Proceedings*, Hamburg, F.R. Germany, 22-26 October 1984.
2. Deb, M.K., *"Wind Energy - "A Source of Power Generation"*, RRB Consultants and Engineers Pvt. Ltd.
3. Freris, L.L., and Others, *European Wind Energy Conference Proceddigns,* Hamburg, F.R. Germany, 22-26 October 1984.
4. John, E., Wade and Steln, Walker, *"Biological Wind Prospecting In Third World Countries"*, National Symposium on Wind Power Development - Opportunities and Strategies, 1990.
5. Karen Conover and Jon Young, *"Improving Wind Power Station Operation and Maintenance*, Tata Mc Graw Hill, New Delhi.
6. Maheshwar Dayal, *"Renewable Environment and Development"*, Konark Publishers (Pvt.) Ltd., 1990.
7. Mani, A., Mooley, D.A., *Wind Energy Resource Survey in India,* Allied publishers, New Delhi, 1990.

CHAPTER 8

Role of Energy Sector in Curtailing Environmental Degradation

K.Nagabhusan Raju
and
Pradeep K. Dadhich
Tata Energy Research Institute,
Bangalore - 560 052

While the rate of energy consumption is certain to grow in the 21st century and beyond, energy production will no doubt cause ecological imbalances, as majority of the energy production will be mainly by burning of fossil fuels. Most combustible sources of energy cannot avoid introducing a large amount of carbon-dioxide and sulphur-dioxide into the atmosphere. As a consequence over a period of time the earth would become uninhabitable because of global warming, green house effect and acid rain.

While discussing the various issues of energy generation and utilisation the chapter throws light on energy conservation, absorption of energy efficient technologies and renewable energy sources.

Introduction

Energy is basic to all aspects of modern society including the defence and security of a nation and the health and vigour of its economy. Today the impact of energy is so extensive that every activity will come to a halt in case of power failure and the economy is thrown out of gear if the energy supplies are disrupted.

Production and consumption of energy give rise to solid, liquid and gaseous waste products which require control to prevent pollution of air and water and interference with the use of land or enjoyment of the natural landscape.

Technical and economic innovations in the field of supply and use of energy will have a fundamental impact on the energy system which is a tightly linked chain composed of four major links: primary sources, energy transformation, distribution and ultimate use. These must be integrated with environmental concerns to provide sustainable development.

Energy Scenario in India

The principal energy sources in India are coal, petroleum products, hydel, nuclear power and biomass. The commercial energy balance for India(1987/88) gives the net availability as 103.35 mtoe with the contribution of coal, petroleum products and power as 40.89%, 42.31% and 12.9% respectively, as against a gross availability of 151.335 mtoe.

During the period 1980-81 to 1987-88 indigeneous production of commerical energy increased from about 72.6 mtoe to over 133 mtoe, at the rate of over 7.8% per annum. majority of this increased production was due to a rise in the use of fossil fuels, basically coal, oil and gas.

Power generating capacity rose from 1713 MW in 1950 to 69000 MW at the end of March 1992; while the electricity generation was 260 billion units during the year 1990-91 rising at an average rate of 10% per annum since 1950. The per capita consumption rose from 14.6 kWh in 1950 to 250 kWh in 1991. Despite this addition in capacity power shortage has continued to hamper normal life and industry. The present expected shortfall in electrical energy is around 9% and peak shortage is around 18%. This includes Thermal, Hydel and nuclear power generation which contribute 71.4%, 26% and 2.6% respectively. 51% of the coal available in India is used for thermal power generation, 44.6% for industrial use and the rest for residential purposes.

Present indigeneous production of crude oil is of the order of 33 million tonnes per annum as against the consumption level of petroleum products amounting to over 54.7 million tonnes per annum. Current projections show a rise in the demand for petroleum products to 101 million tonnes by the end of the decade, by which time indigeneous production is expected to have increased to 58 million tonnes, leaving a deficit of 49 million tonnes.

Sectorial Energy Consumption Pattern

The sectors that have been identified for total available energy consumption are industry, agriculture, transport, residential, other energy uses and non-energy uses. Their relative shares are shown in the figure. As 85% of the energy is consumed by industry, transport and residential sectors, this paper confines its discussion mainly to these sectors.

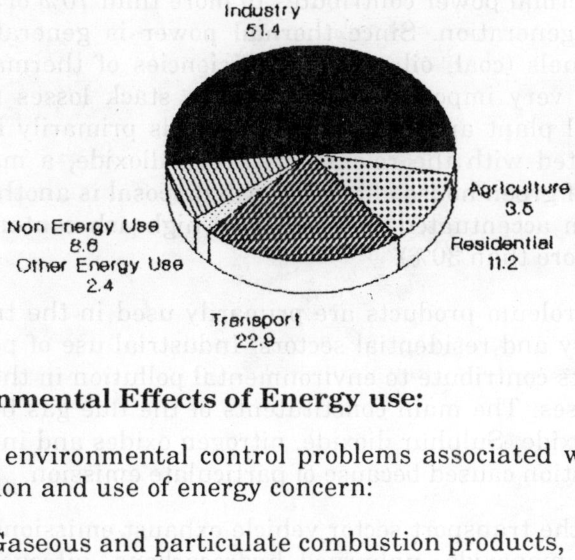

Environmental Effects of Energy use:

The environmental control problems associated with the production and use of energy concern:

- Gaseous and particulate combustion products,
- liquid effluents and solid wastes
- waste heat associated with power station and refinery

cooling water and cooling water plumes,
- damage to visual amenity and enchroacment on landscapes.

The steadily growing use of energy, derived mainly from fossil fuels, is already upsetting the environment, which over the last decade spread over even to regions with less intensive production activities. The environment is one of the most complicated systems known and includes all physical, chemical and biological resources existing on the earth. To fully understand the implications of the energy sector on the environment it is essential to systematically study and assess the inter-relations between these dynamic systems.

Power is generated by burning fuel, splitting the atom, or converting the natural energy of wind, sunlight, falling water or heat from the earth. Each source has its own array of environmental interactions. Without attempting a comprehensive discussion of any here are a few observations about each source of power and its effect on environment.

Thermal power contributes to more than 70% of the total power generation. Since thermal power is generated from fossil fuels (coal, oil etc.) the efficiencies of thermal plants have a very important role. Average stack losses from the thermal plant are around 18% which is primarily heat loss associated with the release of carbon dioxide, a major producer of green house effect. Fly ash disposal is another major problem accentuated in India by a high ash content in our coal (more than 30%).

Petroleum products are primarily used in the transport, industry and residential sectors. Industrial use of petroleum products contribute to environmental pollution in the form of flue gases. The main constitutents of the flue gas being carbon dioxide, Sulphur dioxide, nitrogen oxides and incomplete combustion caused because of particulate emission.

In the transport sector vehicle exhaust emissions include carbon monoxide, unburned hydrocarbons, nitrogen oxides and particulates. This is of major relevance to major cities in India as the number of 2/3 wheelers has increased rapidly in the past decade and is likely to increase in the future. It is estimated that one tanker of petrol consumption produces upto 45 kgs of carbon dioxide adding to green house effect.

Residential sector emissions result from various fuels used for cooking like fuel wood, cowdung cakes, crop residues and kerosene. This affects a large number of people directly exposed to open combustion. Constant exposure to air pollutants generated from open combustion of biomass is likely to cause several respiratory diseases.

Much of the public concern over energy and the environment has focussed on coal whose image has been marred traditionally due to dust, grime, land degradation, mining catastrophes, and chimmney emitting black smoke associated with its use. Coal miners get afflicted with pneumoconiosis (black lung) and it is known that in coal fired power stations the residual fly ash that is left in the open is radioactive.

A study by organisations primarily interested in environmental quality has estimated that the annual environmental impact on land use for a 1000 MW coal fired plant is 3680 hectares if the coal is deep mined and 5650 hectares if it is surface mined. Besides this 65 hectares are required for processing, 895 hectares for transport and 390 hectares for conversion, including 46 hectares for ash storage and 5 hectares for coal storage. In addition this generation unit annually produces 35 lakh cubic feet of ash, 35000 tons of sulphur dioxide and 45 lakh tons of carbon dioxide.

The pollutants which most need to be considered are black smoke, the oxides of sulpur, the oxides of nitrogen, and ash in the form of fine particles. Black smoke is in a special category and will not be considered further because it is both feasible and economic to reduce its emission to very low levels.

Sulphur dioxide and Nitrogen oxides are the major contributers for acid rain. Although concentrations of Sulphur dioxide in stack gases emitted by steam generation plants are usually in the range of 400 to 2000 ppm, the volume of gases produced by the utility the world over is large. Unless this is checked, by the year 2000 it is estimated that the annual emission of so into the atmosphere will exceed 200 million tonnes. It is well known that oxides of nitrogen react with hydro carbons in the presence of oxygen and sunlight to form photochemical smog. Some of the options to reduce acid rain are shown in the diagram below:

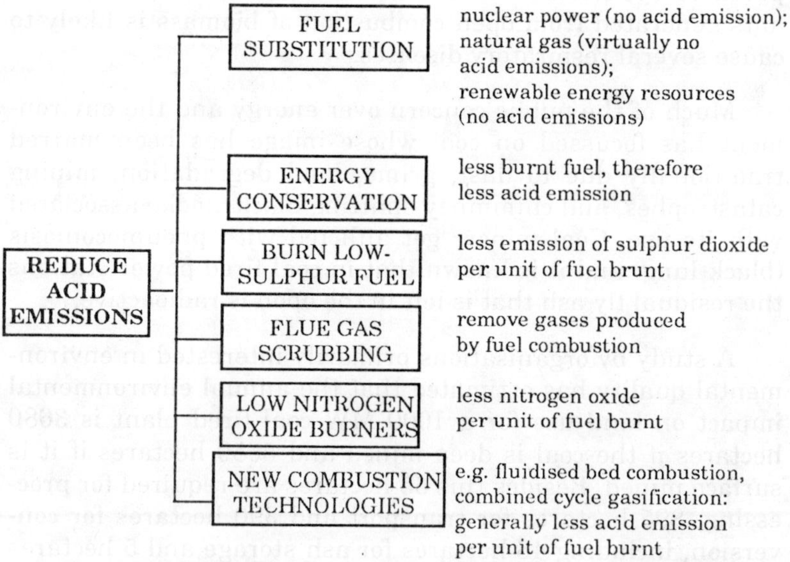

Approximately 90% of carbon monoxide in the air is from natural sources such as decay of organic matter and forest fires. The remaining results from incomplete combustion of fossil fuels, such as oil, coal and gas and other chemical, metallurgical and other combustion processes. Carbon monoxide is also one of the most insiduous of all toxic gases. Being odourless, colourless, imperceptible to the senses, thus giving no sign of its presence. For these reasons carbon monoxide is considered a serious health hazard.

Atmospheric particulate matter is known to arise from several sources : as that emitted from both natural and anthropogenic sources, particles formed by condensation upon cooling of hot gases in the ambient air, particles formed by evaporation of liquid from droplets containing dissolved matter, volcanic eruptions and particles found in the ambient air through reactions converting gases to particulate matters.

Nuclear energy use affects the environment in the form of respiratory cancer to uranium miners and possible release of radioactivity into the environment. Nuclear emissions have been found in water and in the marine life also; but these have to be established that these are at levels which could cause damage to life. There is also fear about the possibility of buried radioactive waste materials getting into water streams and thus contaminating the environment. Nuclear waste disposal problem can be minimised with the proven technology as a 1000 MW plant is likely to produce about 70 cubic feet of high level radioactive vitrified waste.

From the view point of environment hydel power can be beneficial only when naturally occurring falls are available and this cause practically no damage to the environment. But creation of dams for hydel generation can have major problems not only from ecology but also from human angle.

Measures to Mitigate Environmental Degradation

After knowing the impact of energy on environment the question arises as how to curtail the environmental degradation without compromising on the energy generation and utilisation. Energy conservation, use of cleaner fuels, improved technology absorption, waste recycling, and the use of other forms of renewable energy sources, could be possible solutions. the problems are rooted in the economy and society. Given a rational and supportive public policy, Science and technology can provide solutions not only for the 21st century but for ever.

Role of Energy Conservation

While per capita energy consumption in India has been one of the lowest, the energy per unit of GDP is one of the highest even when compared to other developing countries. Infact the energy-GDP ratio in India has been increasing faster than in several other developing countries. For instance,

India is not as efficient a user of energy as some of the other developing countries in Asia.

A major lesson learnt is that energy conservation is a major tangible resource by itself that competes economically with most, if not all, contemporary supply options. In addition it enhances the cost competitiveness among industries, optimises use of our scarce capital intensive energy supply options, protects the environment by curbing pollution, reduces dependence on imported energy.

Energy conservation projects have typically much shorter gestation periods and make our user systems less dependant on utility energy supply. Therefore, energy conservation will have to be an important component of our integrated energy policy.

Let us briefly look at the various technological innovations that help in reducing the energy sector's impact on environment. 'Industrial energy conservation means different things to different people. For some it is the reduction of losses and avoidable waste, to some it is better utilisation of the available energy either by process integration or by process modifications, while for others it may be through effective demand side management.

There are several basic house-keeping energy conservation/efficient measures, that could be adopted.

They are:

- replacement of incandescent lamps by fluorescent or HPSV depending on application,
- high efficiency motors,
- electronic speed control on pumps and fans,
- switch over from V belts to sinchronous belts,
- turning off idling motors, unused lights and fans,
- reducing the peak demand by studying the daily load curve, etc,.

Reducing the energy consumption by these means lessens the demand thereby limiting the power generation and consequential environmental effects.

In the thermal energy utilisation

- improving the efficiency of the equipment,
- reducing stack losses,
- ensuring complete combustion,
- preheating of air and feed to reduce the fuel input etc,. will help reduce pollution levels as well as save energy.

A wide variety of energy efficient equipment are available within the country whose application is likely to bring down the energy consumption levels by 20% on an average. The equipment include: Energy efficient burners, recuperators, regenerators, heat pumps including vapour recompression systems, heat pipes, variable speed drives, etc,. (more details can be had from Teri's book on 'Innovations in India power sector).

Among all types of mechanised freight movement, the railways are the most attractive, being seven to eight times more fuel efficient than road transportation. Transportation strategies should therefore involve the long distance movement of goods by the railways and those to nearby supply points by road transport. It is also worthwhile to consider use of coal at the pit head thermal power stations or in the immediate vicinity. Because of the cheaper cost of transportation by water, it is prudent to combine road and rail transport with water transport.

Interfuel substitution have shown good results and its commercialisation has to be accelerated. Compressed Natural Gas (CNG) and methanol should be available for commercial use by the end of this century.

Energy conservation in the residential sector through recycling of waste, improved chulhas, biogas plants and the other renewable sources of energy is probably the important way to reduce the effect on environment.

Energy efficient technologies

There are several energy efficient technologies which have been successfully demonstrated in India and abroad. Only an outline of a few technologies is given below, as discussion in detail is beyond the scope of this chapter.

(a) *Hydrocarbon vapour recovery plant:* The plant recovers the petrol vapours that are currently lost into the atmosphere during the loading of tank trucks in fuel storage depots. Vapour emission during the loading of tank trucks can reach very high levels, because loading work is concentrated in short periods of a few hours per day. This recovery plant provides an acceptable solution for absorbing emission points at a reasonable cost, and thus it is justified in terms of economy and environmental hygiene as well as in terms of energy savings.

(b) *Isobaric Double Recycle (IDR)- a new Urea synthesis process:* This process is characterised by a high conversion per pass through the synthesis reactor, an easy isobaric recycle of the unreacted matter to the reactor and a relatively low steam consumption; steam consumption a range of 40% less than the one required for the present technologies. Reducing the steam consumption means reducing the burning of fossil fuels and associated pollution.

(c) *To use Colliery methane as a new energy source:* To use mines gas which is a waste product of coal mining operations in the industries surrounding the collieries.

(d) *Incineration of solid toxic waste for thermal applications:* The technology can be used to incinerate composting rejects, rejects from sorting and methanisation processes, various types of plastic wastes, synthetic waste, tyres etc,. The emission levels from the stack are proved to be well within the limits.

(e) *Methanisation of Sugar-cane molasses stillage:* The process converts most of the soluble carbonated pollutants into methane and carbon dioxide. This biogas is then beneficial to some distillaries then others depending on such factors as local energy costs, national pollution prevention standards and whether or not bagasse is used. For example, a molasses distillary discharging wash with a Chemical Oxygen Demand (COD) of 30 tonnes and operating 300 days a year, produce 2 million Nm^3 of methane per year or 1750 tonnes of oil equivalent.

(f) *Congeneration:* Congeneration is the means to recover waste heat and has a very high efficiency of energy utilisation. This is an effective energy conservation technique with the sequential production of electrical or mechanical power and process heat from the primary fuel. Utilisation of cogeneration processes improves the energy efficiency to around 80-90%. Although cogeneration is a well established technique it has not been exploited by the indian industry fully.

Four consideration play an important role in these technologies: (1) the reduction in fossil fuel consumption (2) the reduction in the volume of waste arising (3) the contribution to solving the environmental problem and (4) the ease of incorporation into the existing heat and power system.

Progress in converting to energy saving technologies has so far been much lower than it should be. Because of the important of the industry's contribution to national development, both economic and social, its ability to minimise energy consumption and the consequential environmental effects in the transition will have profound effect on all of us.

The other methods of reducing pollution of the atmosphere apart from all the above methods are:

— to select fuels of low pollutant characteristics for regions of severe pollution,

— to remove the pollutant from the fuel before combustion.

— to remove the pollutant from the gases after combustion,

— to use tall chimneys.

Role of Renewable energy sources

Renewable energy is the form of energy that is continuously available for use. However certain renewable sources such as forests can be depleted when over-exploited. It is known for its reduced effect on environment. This is one thrust area in which the country should gear up for implementation, as plenty of proven technology is available in the country. A few of these technologies are briefly discussed without going into technical details.

Solar energy systems use natural, non-mechanical forces including sunshine, shading, and breezes to help heat and cool homes with little or no use of electrical controls, pumps and fans. Other solar energy technologies are solar collectors, photovoltaics and solar pond etc. The potential availability of wind energy is estimated to be of the order of 50000 MW, only along the coastal regions on the basis of land availability. Wind energy systems can also be off-grid and can be used frequently in remote sites. These are commonly used to recharge batteries and to pump water.

Organic solid wastes and residues from biomass are now generally recognised as renewable fuels, which must increasingly supplement fossil fuels. However utilisation of such wastes and residues requires some modifications in the equipment, many of which have been developed and tasted within the country. A vast number of gasifiers are now available in various sizes both for domestic and industrial purposes.

Renewable energy sources have attracted considerable interest over the past decade promise to play a growing role in the future energy picture. The steel, aluminium and glass panels required for harnessing solar and wind energy have to be replaced with less energy intensive material. Short term cost dominated by R & D expenditure and high front-end investments, do not compare favourably with the cost of conventional technologies without subsidies., In the long run, however, advancement in technology, mass production and expected increases in the cost of fossil fuels and their effect on environment, will undoubtedly improve the cost effectiveness of revewables.

Conclusion

In spite of all these aspects the role of energy can be best viewed from the fact that we have seen coal plants converted to oil to preserve air quality; oil plants converted to coal because of concerns about imports; nuclear plants first promoted then discouraged. All these activities serve to illustrate the vascillating policies that a diversity of fuel types is the best insurance against surprises.

Any change in the energy policy should be directed to follow the approach that makes economic, and social sense. Efficiency management, institutional changes and the adoption of a variety of energy efficient technologies using are options, but the real question is in implementing it.

Finally, In our quest for optimum solutions we should keep our options open and continue with a diverse mix, further developing each of the technologies and lessening their impact on environment. We must not overlook the fact, that we need substantially more power as well as increasingly successful conservation practices to fulfill the needs of 21st century.

REFERENCES

1. TERI Energy Data Directory and Yearbook (TEDDY), 1989.
2. European Community demonstration projects for energy saving : Nos. 4, 11, 14, 89.
3. Energy for our Globe's people - William S Lee, ENVIRONMENT, Vol.32, No.7.
4. Energy Policy for India - Dr. R.K. Pachauri.
5. Efficient use of Energy - I.G.C. Dryden.
6. Energy and Environment cost benefit analysis - R.A. Karam, K.S. Morgan
7. Environment, Resources and Conservation - Susan Owens and Peter l Owens.
8. Energy Technology handwork - Considine.
9. Innovations in the India Power Sector - Leena Srivastava, Dr. R.K. Pachauri.
10. Fundamentals of Energy Engineering - Albert Thuman
11. Standard handbook of Electrical Engineers - Donald G Fink.
12. The MNC's cast their net on power sector
13. Energy Conservation: Global perspectives - Dr. R.K Pachauri.
14. Oil conservation opportunities in transport sector - K.N. Venkatasubramanian. (12, 13, 14 - articles in THE HINDU).

CHAPTER 9

Geothermal Energy Exploitation, Utilisation and Worldwide Development

Dr. Dharmapuram Venkat Reddy
*Karnataka Regional Engineering College, Surathkal, Mangalore - 574 157**

The earth is a hot plastic mass of rock covered by a thin veneer of cooled rigid crust. Most of the exploitable geothermal resources are hydrothermal in nature, and reserves are located in natural water-filled rock aquifers with anomalous heat. Reservoirs that produce high temperatures of 100°C and above are considered to be high-enthalpy reserves and are mainly utilized for power generation. Those producing 20-100°C temperatures are low-enthalpy deposits, and are utilized for space heating and for applications in agriculture, aquaculture, food processing, etc. Experimental results have verified that anomalous hot dry rock energy potential exists and is mostly confined to the igneous rocks. Such rocks cover 25% of the global continental area. Efforts are on to utilize hot dry rock technology for commercial use. Recent studies confirmed that worldwide geothermal electrical power capacity increased to 5828 MW, which means an increase of 50% in just five years. It is predicted that in the next two decades utilization of geothermal energy will expand in the high-temperature zones of the world. For long-term planning, geother-

---

* Reproduced, with permission from "Energy Environment Monitor" vol. 7(2) September, 1991.

mal energy must rank as one of the world's major alternative fuel energy resource.

Introduction

In 1913, the first attempt was made to extract the natural heat contained within the earth in the form of geothermal steam. Since then much progress has been made and today, there is a thriving geothermal industry. It is believed, however, that geothermal energy as a natural resource is still much neglected and that a far greater potential exists than is generally recognized (Bichan 1988). Geothermal heat is gradually being used as a non-conventional energy source in view of the ever-increasing demand for power and the limited potential of the conventional sources. Direct or non-electric use of geothermal energy is increasing steadily. At present, worldwide geothermal electrical power capacity stands at 5828 MW representing a 50% increase in the past five years. Most geothermal potential is confined to volcanic and tectonic areas (Nielson 1991).

In this chapter an attempt has been made to study the sources of geothermal energy, status of worldwide geothermal energy development, utilization of high-and lowthermal fluids, exploitation and working technology of hot dry rocks (HDR) prospects for geothermal fields in India, and environmental benefits.

Sources of geothermal energy

The earth is a reservoir of natural heat contained in the form off geothermal steam and acts as a great heat engine during episodes of volcanic eruption. Hot springs (geysers) and fumaroles are other surface manifestations of the earth's heat content. It has been recorded that at 10 km depth the average temperature is 200-350°C and at 30-35 km it is as high as 500-750°C. These experimental findings concluded that earth is a high-grade thermal resource and that geothermal energy can be exracted from its depths comparatively near the crust's surface. The major geothermal and volcanic belts are situated in the circum-Pacific margin, island groups

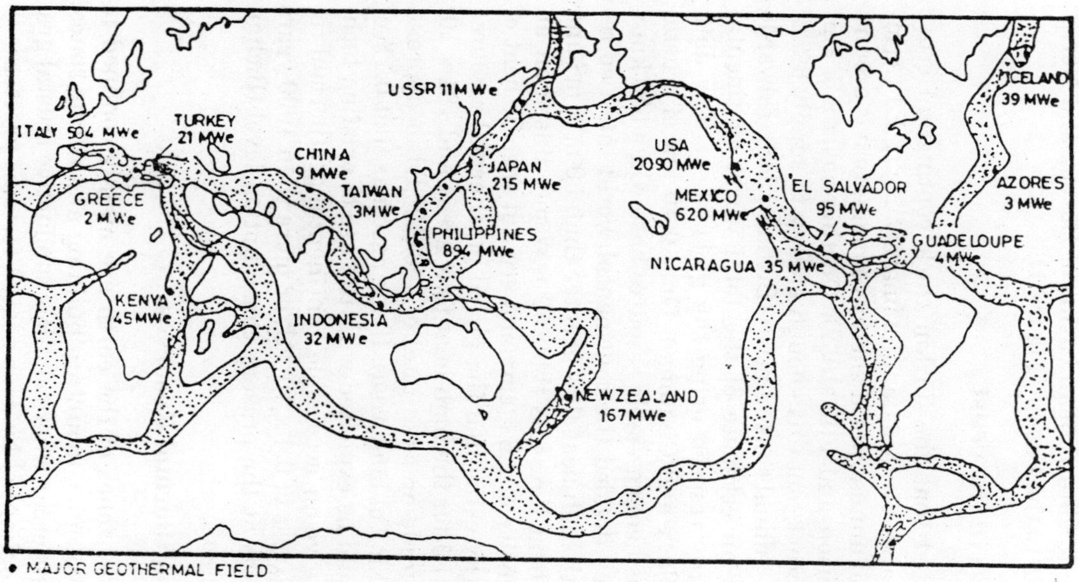

Fig. 1 Distribution of world's high temperature belts showing installed geothermal electrical generating capacity. (After H R Bichan 1988).

of the mid-Atlantic rift, rift zones of East Africa and adjacent Middle East and irregular belts of mountains and basins extending from the Mediterranean basins of Europe and North Africa across Asia to the Pacific (Figure 1).

Energy reserves in the crust

It is estimated that the 35 km thick continental crust covering 1 km^2 area contains approximately 26.31 EJ (Armstead 1987). Total annual world energy consumption is around 370 EJ (Stephenson and Fllis 1987). On this basis a 1 km^2 surface area of crust contains enough energy to supply 6% of the world's total annual consumption (Bichan 1988). Edwards *et al.* presented an estimate of the total world continental geothermal energy reserve using the global heat flow data compiled by Pollack and Chapman. Only 0-10 km of crustal depth is taken as being a viable resource base. On a 550 km x 550 km grid they divided the continental world into two regions - areas with high heat flow greater than 50 mW m^{-2} and those with low heat flow less than 50 mW m^{-2}. Based on an average rock density of 2.5 g cm^{-3} and crustal specific heat of 770 J kg^{-1} $°C^{-1}$, thy estimated the global geothermal resource base as 245×10^6 EJ for the high heat flow areas and 181×10^6 EJ for the low grade areas (a total of 426×10^6 EJ or approximately 2.95 EJ km^{-2} of land area). If one assumes that 0.1% of this resource could be exploited with the use of existing technology it would give an 'accessible' continental geothermal resource base in excess of 0.4×10^6 EJ, or more than 1000 years of energy reserves at the present consumption level (Bichan 1988).

Exploitable Geothermal Reserves

The heat-flow zones of the earth are mainly confined to the magmatic, volcanic, mountain building and rifting zones. Geothermal spots occur where shallow conductive crustal gradient has been enhanced by deep-seated tectonic processes in the earth. Exploitable geothermal reserves are located in water-filled aquifers with irregular heat.

Geothermal reserves are classified on the basis of presence or absence of natural water-geothermal aquifers where water occurs in association with natural crustal thermal heat and those where HDR are situated without water. Exploitation of hot water zones in thermal aquifers, has reached the commercialization stage whereas energy from HDR zones is still in the experimental stage.

Geothermal aquifers are further classified on the basis of fluid temperatures, which vary from 20°C to 350 °C. Geothermal fluids with temperatures of 100 °C and above are high temperature or high-enthalpy deposits. Reservoirs that produce hot fluids in the range of 20-100 °C are considered as low temperature or low-enthalpy deposits. High temperature reservoirs are utilized for production of the power, whereas low-temperature reservoirs are utilized by various industries for process heat based on the temperature gradient (Figure 2).

High-enthalpy geothermal aquifers

A typical cross-section of a high temperature geothermal aquifer is shown in Figure 2. Volcanic formations where high temperature aquifers are associated with rock formations of basalts, rhyolites and andesites are potential zones. Exploitation of these aquifers needs detailed integrated surveys such as geological, hydrogeological, geochemical and geophysical. Based on these surveys, systematic thermal gardient analyses are interpreted for further exploitation of these hot reservoirs for power generation. The hot fluids that are pumped out from the aquifer are released into electricity generating turbines. Excess hot fluids are injected into the reservoir through injection wells. Schematic diagram of occurrence and exploitation of high temperature geothermal aquifers is shown in Figure 3.

Low-enthalpy reserves

Aquifers with water temperatures less than 100°C are generally classified as low temperature aquifers. These are abundant in nature when compared with high-enthaply aqui-

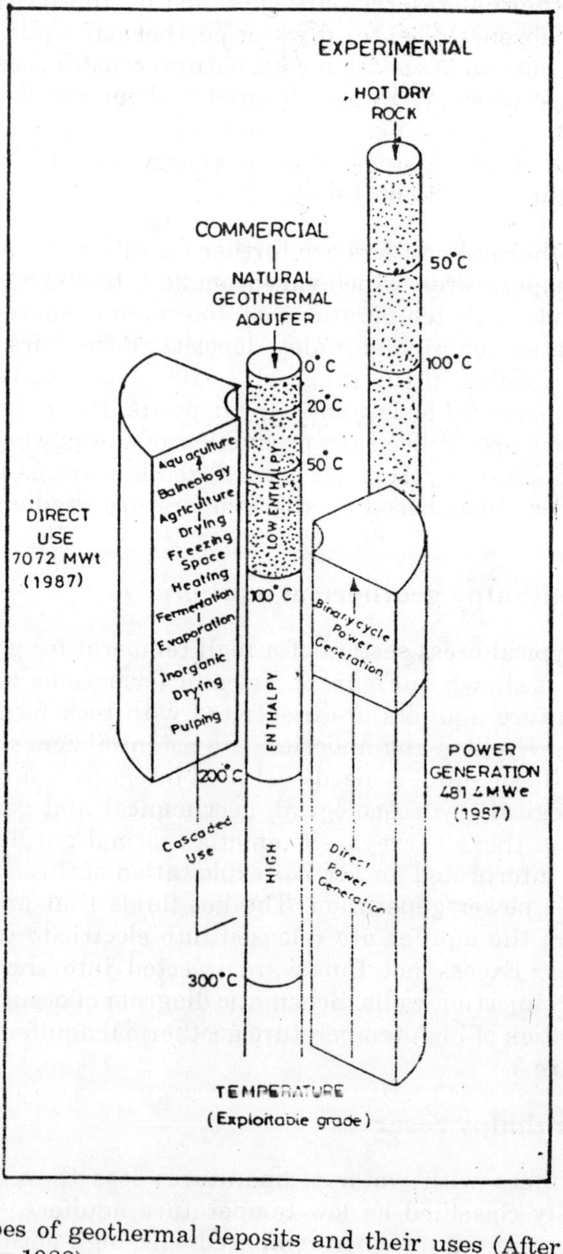

Fig. 2 Types of geothermal deposits and their uses (After HR Bichan 1988)

fers. Low-temperature aquifers are mainly located in sedimentary basins. It has been estimated that more than 30% of global energy consumption is at temperatures below 80 °C (Bichan 1988).

Low-temperature geothermal energy is utilized for space heating, and for industrial and agricultural applications. In Iceland at Reykjavik, Reykir, Olafsfyoudur, Selfoss and several other places low thermal energy is utilized for heating houses, public buildings, swimming pools, etc.

At present 35 countries utilize 11605 MW geothermal energy for non-electric applications, whereas 18 countries utilize 5 657 MW for electric applications (Table 1). Construction of 502 MW capacity plants are underway in Dominica, Costa Rica, Guatemala, St.Lucia, and many countries are planning to increase use of geothermal energy.

There is a misconception in India that geothermal energy can only be utilized for power generation. In countries where power generation is not possible because of low geothermal energy, thermal waters are being utilized for heating, cooling, industrial and agricultural purposes (Kakar, Handa 1988).

Exploitation of hot dry rocks (HDR)

Exploitation of HDR technology is still at the experimental stage. It is planned to utilize HDR energy for commercial purposes. The term HDR is normally used for areas of the crust that exhibit anomalous conductive heat flow (general thermal gradient more than 25°Ckm^{-1}). The heat flow is not asscoiated with water. More than 25% of the earth's area is covered by crystalline hot dry rocks, which are the source of HDR energy, Igneous and metamorphic terrains are the principal sources.

Working technology of HDR

Based on geological data; a suitable area is selected. Two wells are drilled up to 3000-2500m depth. One well is for injection of water from the surface, another for extraction of hot water from the artificial aquifers. The objective of this

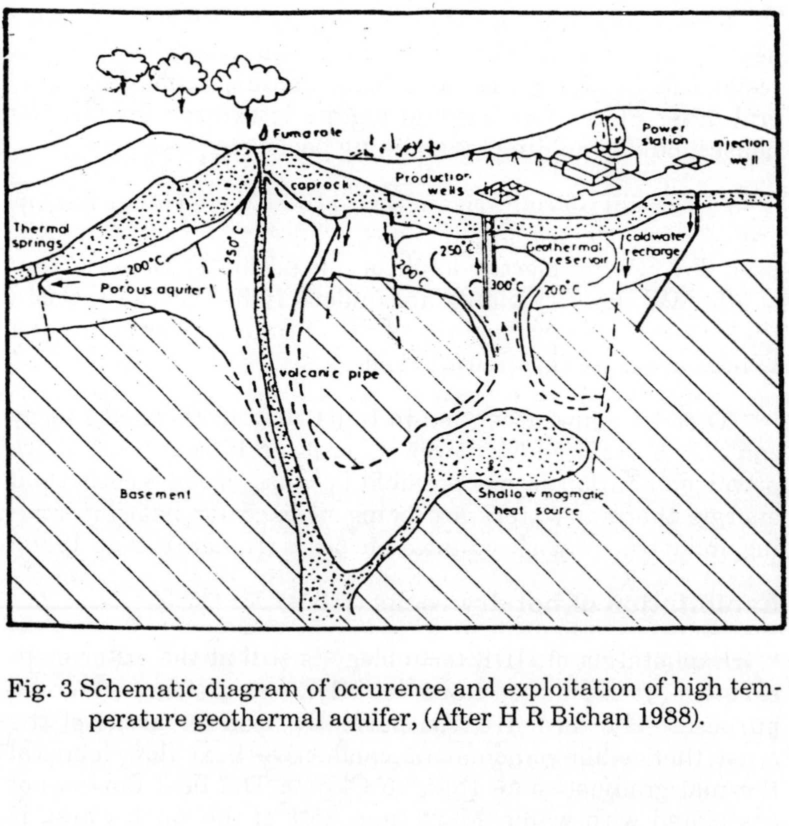

Fig. 3 Schematic diagram of occurence and exploitation of high temperature geothermal aquifer, (After H R Bichan 1988).

technology is to create an artificial permeable aquifer reservoir in the crust by hydraulic fracturing. Through the injected well, water is pumped out from the production well. Hot water is supplied to the generating station for power generation. But this technology leads to a huge loss of water. To overcome the problem viscous gel is injected into the artificial aquifer (Figure 4.).

Extensive research is on utilize HDR energy, particularly in USA, Japan, France, Germany, Sweeden and USSR. In future if HDR technology can overcome its drawbacks, it will be possible to exploit igneous metamorphic terrains for energy.

Prospects of geothermal fields in India

India has vast potential for geothermal energy with more then 300 hot water springs all over the country. These hot springs are situated mostly in Bombay, Ratnagiri, Ladakh, Pugga, Manikaran, Badrinath, Gaurikund, Tapoban, Rajgir, Agnikund, Brahmkund, Surajkund, Gundala and Puttur. The geothermal wing of Geological Survey of India (GSI) has undertaken to exploit geothermal energy from Pugga thermal spring in Ladakh, Kasol in Parbati valley, kalath and Bashist in Beas Valley, Tattapani on the banks of Sutlej river, and other possible hot springs.

Table 1
Worldwide geothermal capacity (MW)

	Use	
Country	*Non-electric*	*Electric*
1	*2*	*3*
Algeria	13	
Australia	11	
Austria	4	
Azores		3
Belgium	93	

1	*2*	*3*	*4*
Bulgaria	293		
Canada	2		
China	2143	20.9	
Colombia	12		
Costa Rica			55*
Czechoslovakia	105		
Denmark	1		
El Salvador		95	+ 20*
Ethiopia	38		
France	337	4.2	
Germany	8		
Greece	18	2	
Gautemala	10		
Hungary	1276		
Iceland	774	44.6	
Indonesia	142.3		
Italy	329	545	+ 160*
Japan	3321	214.6	+ 55*
Kenya	45		
Mexico	28	700	+ 50*
New Zealand	258	293	+ 2*
Nicaragua		35	
Philippines	1	894	+ 110*
Poland	9		
Romania	251		
Switzerland	23		
Taiwan	11		
Thailand	2	0.3	
Tunisia	90		
Turkey	246	20.6	
UK	2		
USA	463	2770	
USSR	1133	11	+ 50*
Yugoslavia	113		
West Indies		4.20	
Total	11605	5657.4	+ 502*

*Under construction.

Source: Hutter 1990, Freeston 1990.

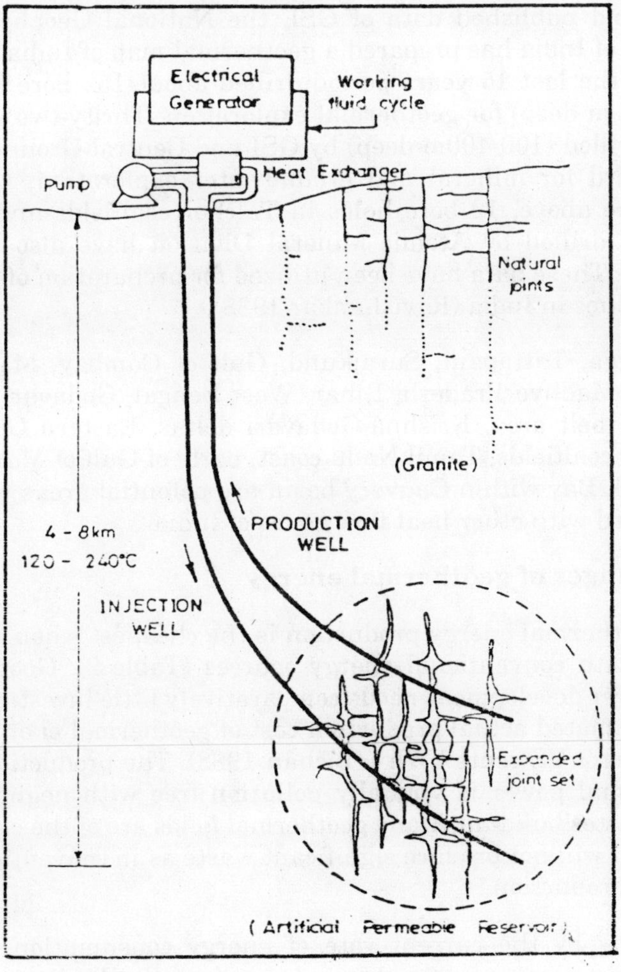

Fig. 1 Conceptual diagram of a hot dry rock geothermal system (After H R Bichan 1988).

The GSI has taken up extensive exploration work in India to study the geothermal heat flow patterns. Based on field work and published data of GSI, the National Geophysical Survey of India has prepared a geothermal map of India. GSI during the last 15 years period drilled about 152 bore holes (40-700 m deep) for geothermal exploration. Thrity-two wells were drilled (100-400m deep) by GSI and Central Groundwater Board for mineral and groundwater exploration. Apart from the above, 12 bore holes in Talcher coalfields and another 2 drilled by Atomic Mineral Division have also been studied. These data have been utilized for preparation of heat flow-values in India (Ravishankar 1988).

Pugga, Tattapani, Surajkund, Gulf of Cambay, Maharashtra, Madhya Pradesh, Bihar, West Bengal, Godavari valley coal belt area, Krishna-Godavari deltas, Eastern Ghats, Talcher coalfields, Tamil Nadu coast, parts of Gulf of Mannar and palk Bay within Cauvery basin are potential areas when compared with other heat flow zones of India.

Advantages of geothermal energy

Geothermal energy production is the cheapest when compared with conventional energy sources (Table 2). Geothermal power development needs comparatively little investment. It is estimated actual production cost of geothermal energy is as low as 2 US cents kWh^{-1} (Bichan 1988). The production of geothermal power is virtually pollution free with negligible toxic wastes build-up. Many geothermal fields are of the closed type and will not produce significant waste as in conventional energy production.

Going by the current rate of energy consumption, the total proven reserves of gothermal energy, in crude oil equivalent, is about 7 600EJ which would suffice for 60 years. The total reserves of coal estimated at 135 410 EJ will last 1 000 years. Nuclear fuels are obvious choice but long-term utilization is heavily dependent on development of breeder reactors. Total uranium deposits of the world are estimated at 2490 EJ. Extensive nuclear development may also not possible because of the growing anti-nuclear environmental lobby. In the long term, geothermal energy must rank as one of the world's major alternative fuel resources (Bichan 1988).

Conclusions

1. The earth is a large reservoir of natural heat energy, but the potential has not been explored satisfactorily.
2. It is estimated that in the next two decades both high-enthalpy and low-enthalpy reservoirs and allied geothermal industries will expand rapidly.
3. It is estimated that 1 km^2 of crust contatins enough energy to supply 6% of the world's total annual consumption. The global annual energy consumpion is about 370 EJ.
4. Heat flow zones in the earth are mainly confined to magmatic, volcanic, mountain building and rifting zones. Dormant volcanoes will become active depending upon the pressure developed in the crustal layers of the earth.

   On 3 June, 1991, the dormant Mount Unzen (Japan) erupted after 200 years and Mount Pinatubo, Philippines erupted after a dormant period of 600 years on 12 June, 1991. This clearly indicates that volcanic zones are sources of geothermal energy.
5. High-enthalpy aquifers are utilized for power generation whereas low-enthalpy reservoirs are explored for various applications based on thermal gradient.
6. Extensive research is on for exploitation and utilization of HDR energy. Igneous and metamorphic terrains are the potential sources of this energy. HDR does not need any natural thermal aquifers. Energy from dry rocks can be exploited directly.
7. India has a vast potential for geothermal energy. More than 300 hot springs are situated in the country.
8. Geothermal energy production is environmentally safe and its development need is little investment as compared with other conventional energy sources.

Acknowledgements

The author is grateful to Prof. N.R Shetty, Principal, Karnataka Regional Engineering College, Surathkal for providing necessary facilities for the study. The author is highly indebted to Prof. S S Bhavikatti, Head Department of Civil Engineering for helping in various stages of the survey.

REFERENCES

1. Armstead H C D, Tester J W. (1987), *Heat Mining a New Source of Energy,* London: E&FN Spon.
2. Bichan H R(1988), *Geothermal Energy Fuel for the Future,* London: The Institution of Mining and Metallurgy.
3. Edwards L M(1982), *Handbook of Geothermal Energy,* Houston: Gulf Publishing Company.
4. Freeston D H (1990), Direct uses of geothermal energy in 1990, *Geothermal Resources Council Bulletin 19(7):*188-198.
5. Hutter G W(1990), Geothermal electric power-a 1990 world status update *Geothermal Resources Council Bulletin* 19(7):175-187.
6. Kakar Y P, Handa B K (1988) Geothermal water in Himachal Pradesh, *BHU-JAL Journal* 3(4):1-8.
7. Nielson D L (1991), Geothermal energy, the status of worldwide geothermal development, *AGID News* 65: 17-18.
8. Pollack HN, Chapman DS(1977) On the regional variation of hear flow, geotherms and lithospheric thickness, *Tectonophysics* 38:279-296.
9. Ravisankar (1988) Heat flow map of India and discussions on its geological and economic significance, *Journal of Indian Minerals* 42(2):89-110.
10. Stephenson B, Ellis J H M(1987)World energy consumption and reserves, Robertson Research Geothermal Division:10. [Working paper E 87-2]
11. Venkat Reddy D (1989), Geothermal energy - future resource, In *Proceedings of the Non-conventional Energy Sources Perspectives for the Year 2000 AD*: 18-19, Institution of Engineers (India), Mangalore and Manipal Institute of Technology, Manipal.

Table 2

Levelized costs for new baseload electrical generating capacity, 1986 ($ US)

Resource type	*Plant size (MWe)*	*Power Plant capacity costs ($kw⁻¹)*	*Construction time (years)*	*Annual Power plant costs* (cents kwh)*	*Total generating costs (cents kwh⁻¹)*
Geothermal	110	750	2-3	4.8	4.5
Coal	1200	1300	6-8	8.1	6.1
Nuclear	1200	2650	10-15	6.4	7.5
Hydro	1200	1875	8-12	4.5	4.6

*Based on 17% annual fixed rate; 80% load factor (7 000 h $year^{-1}$ at capacity): coal cost of $ $40t^{-1}$.

Selected Bibliography

1. Al Alm (1989), "Energy Economics and Environmental Policy", Environmental Science & Technology,.Vol.23(6).

2. BW Ang (1989), "Modelling for Petrol Consumption under Data Constraints: the case of Singapore",. PAJE, 3(2) DEC

3. HCD Armstead and JW Tester (1987), Heat Mining a new source of Energy ,. E & FN SPON, London

4. SK Baksi (1988), "Achievement in Cost Reduction through Planning of Energy Conversation in Durgapur S.P, Energy Mgmt., 12(1)

5. Bansal (1990), Energy Options for the 90s, Tata McGraw Hill, NewDelhi

6. A Barnett, B Martin & Hoffman, (1982), Rural Energy and the Third World, Porgamon, New York

7. JC Bergstorm (1990), Concepts and Measures of Economics Value of Environmental Quality: A Review, J/o Emt. Mgmt, 31(3) Oct

8. HJ Bhabha & NB Prasad(1958), A study of the contribution of atomic energy to a power programme in India, Proc., 2nd.UN Conf. on AE, Vol.1

9. Ramesh Bhatia & Samantra (1989), "Wind Electricity Generation in India: Performance evaluation and economic analysis", PAJE, 3(1) JUNE

10. AP Bhatnagar & BS Panesar (1989), "Energy Conservation in Agricultural production and Agro Based Industries", ENERGY MGMT., Vol. 4

11. AP Bhatnagar and BS Panesar (1989), "Energy Consumption and Conservation Potential in Irrigation & Agro-based Industries", PAJE, 3(1) JUNE

12. HR Bichan (1988), Geothermal Energy: Fuel for the Future, IMM London

13. RE Chatfied, Hein & Moyer (1990), Long term Earnings in the Electric Utility Industry: Accuracy & valuation Implication, Financial Review, 25(3)

14. Karen Conover & J Young., Improving Wind Power Station Operation and Maintenance,. Tata McGraw Hill, New Delhi

15. E Cook (1976), Man, Energy, Society, WH Freeman USA.,

16. J Darmstadter et al (1971), Energy in the World Economy: A Statistical Review of Trends in Output, Trade & Consumption, John Hopking Press (US).

17. Dayal (1990), Renewable Environment and Development, Konark Publishers,

18. PN Dhar & DU Sastry (1967), Demand for Energy in North -West India, Asia (Delhi),

19. PJ Dhrymes & M Kurz (1964), "Technology and Scale in Electricity Generations", Econometrica, 32, July

20. Edwards (1982), Handbook of Geothermal Energy, Gulf Publ Co., Houston

21. F Felix (1964), Growth of Energy Consumption and National Income Throughout the World,.IEEE Spectrum, July.

22. DH Freeston(1990), Direct Uses of Geothermal Energy in 1990, Geothermal Resources Bulletin, 19(7), p188-98,

23. Arun Ghosh (1991), "Eighth Plan, Challanges & Opportunities-VI; ENERGY:Coal, Electricity Energy Cons.",.Eco.& Pol.Weekly,XXVI(13)

24. J Goldemberg et al (1987), Energy for Development,. Oxford IBH,

25. RS Goud(1989),"Rural Electrification through Cooperatives", Yojana, 33(17)

26. Hariharan, Power Systems for the year 2000 and beyond, Tata McGraw Hill, NewDelhi

27. Jeffrey Jarrett (1990), Forecasting Electric Power & Gas Production Data: A comparison of methods,. Business Journal, 7(1)

28. GK Kadekodi and Anup Sharma (1988), "Cost of Coal Extraction: A lng run View for India", PAJE, II(2)DEC.

29. C Kashkari (1975), Energy Resources, Demand & Conservations with special reference to India,. Tata McGraw Hill, New Delhi

30. A Mani & DA Mooley(1990), Wind Energy Resources Survey in India, Allied Publishers New Delhi.

31. GK Mitra & MS Reddy (1985), "Cost and benefit of Rural Electrification: A case study of A.P.", PRAJNAN, XIV-1

32. H Mohr (1987), "How Variable are Alternative Energy Sources ?", Sulzer Technical review, Vol.4,4-9.

33. Muralidhar and Tretter (1991), Impact of Special Requirement on the Estimation of Electrical Demand, Management Sc.,37(1) Mar

34. Nagrath & Kothari, Modern Power System Analysis 2/e, Tata McGraw Hill, New Delhi
35. RK Pachauri(1987), Global Energy Interactions (Proceedings, 5th International Conference of IAEE).

36. RK Pachauri(1977), Energy and Economic Development in India, Praeger, New York

37. JK Parikh(1980), Energy Systems and Development, Oxford Univ. Press, NewDelhi

38. KS Pari (1976), Energy, second India studies for Ford Foundation, Macmillan, NewDelhi

39. PJG Pearson & PJ Stevens (1987), :Integrated Energy Forcasting and policy in LDCs: Problems and prospects", PAJE, 1(i) JAN.

40. Sultan H. Rahman(1988), :Aggregate Energy Demand Projection for India: An econometric approach", PAJE, II(1) JULY

41. S. Ramesh and T.V. Natarajan(1987), "Policy Options in rural Electrification in India: With special reference to pumsets, PAJE,1(1)JAN.

42. V. Ranganatham (1989), "Impact of Rural Electrification in Developing Countries: Special reference to India", URJA, 25(3)

43. PP Rao (1990), "A Dynamic Programming-approach to Determine Optimal Man-power Recruitment Policies", J/O OR Society, 41(10)

44. B Sudhakar Reddy (1989), "Rural Energy Systems for 2000 AD", Energy Mgmt. Vol.13(3)

45. AK Reddy et al (1991), "A Developing Focussed End-Use Oriented Electricity Scenario for Karnataka" INDIAN ECO.Rev.,25(1)

46. Joga Singh (1990), "Production Planning for Thermal Elec. Power Industry of Eastern India", INDIAN Eco.Rev., 25(1)

47. C Starr(1979), Current Issues in Energy, Pergamon Press,

48. Dr. Subroto (1989), "Long-term Prospects for Oil Market Stability", PAJE, 3 (1) JUNE.

49. Mr Suresh, BN Misre & V. Joshi(1990), "Harinagar Rural Energy development project" (DIAGNOSTIC), VIKALPA, 15(1)

50. DD Tewary & VM Rao (1989), Modelling Energy use in Indian Agriculture Sectoral Approach", VIKALPA, 14(4)

51. WE Tyner(1978), "Energy Resources and Economic development in India", Allied Publishers.

52. B Viswanath (1991), "Environment Audit: Looking beyond its scope",, Fortune India, Oct.15